DELICIOUS

Delicious

THE EVOLUTION OF FLAVOR AND
HOW IT MADE US HUMAN

Rob Dunn and Monica Sanchez

PRINCETON UNIVERSITY PRESS

PRINCETON AND OXFORD

Published by Princeton University Press
41 William Street, Princeton, New Jersey 08540
99 Banbury Road, Oxford OX2 6JX

press.princeton.edu

First paperback printing, 2022
Paperback ISBN 9780691242088

The Library of Congress has cataloged the cloth edition as follows:

Names: Dunn, Rob R., author. | Sanchez, Monica (Anthropologist)
    author.
Title: Delicious : the evolution of flavor and how it made us human /
    Rob Dunn and Monica Sanchez.
Description: Princeton : Princeton University Press, [2021] | Includes
    bibliographical references and index.
Identifiers: LCCN 2020037562 (print) | LCCN 2020037563 (ebook) |
    ISBN 9780691199474 (hardcover) | ISBN 9780691218342 (ebook)
Subjects: LCSH: Taste—Physiological aspects. | Smell—Physiological
    aspects. | Perception—Physiological aspects. | Flavor.
Classification: LCC QP456 .D86 2021 (print) | LCC QP456 (ebook) |
    DDC 612.8/7—dc23
LC record available at https://lccn.loc.gov/2020037562
LC ebook record available at https://lccn.loc.gov/2020037563

British Library Cataloging-in-Publication Data is available

Editorial: Alison Kalett, Whitney Rauenhorst
Production Editorial: Terri O'Prey
Text Design: Karl Spurzem
Jacket/Cover Design: Jessica Massabrook
Production: Jacqueline Poirier
Publicity: Sara Henning-Stout, Kate Farquhar-Thomson
Copyeditor: Jodi Beder

Jacket/Cover illustrations by Natalya Balnova / Marlena Agency

This book has been composed in Arno Pro with Futura display

Printed in the United States of America

Why do we eat?
In order to pursue the flavor of things.

—HSIANG JU LIN AND TSUIFENG LIN[1]

CONTENTS

Prologue: Eco-Evolutionary Gastronomy   ix

CHAPTER 1. Tongue-Tied   1

CHAPTER 2. The Flavor-Seekers   25

CHAPTER 3. A Nose for Flavor   53

CHAPTER 4. Culinary Extinction   80

CHAPTER 5. Forbidden Fruits   114

CHAPTER 6. On the Origin of Spices   129

CHAPTER 7. Cheesy Horse and Sour Beer   154

CHAPTER 8. The Art of Cheese   182

CHAPTER 9. Dinner Makes Us Human   203

Notes   215
References   245
Illustration Credits   261
Index   263

PROLOGUE: ECO-EVOLUTIONARY GASTRONOMY

The human craving for flavor has been a largely unacknowledged and unexamined force in history.

—ERIC SCHLOSSER, *FAST FOOD NATION*

A number of years ago, on a path on top of our favorite island in Croatia, we stumbled upon a series of abandoned structures. Later it would become clear that they were stone pens in which people once kept sheep. The structures were circular and immense, and in amongst them we also found the remains of what appeared to be a house once inhabited by a family. These ruins may well have been thousands of years old. The island was long inhabited by Illyrian pastoralists. It has been argued that these pastoralists were the inspiration for the Cyclopes in Homer's Odyssey. They slept in stone houses or caves and lived lives dependent upon sheep, the milk of sheep, the meat of sheep, and even the wool of sheep. The structures we found might have been Illyrian. Or they might have been far more recent. The island is a place in which ancient structures and newer ones comingle easily in ways that are not always legible. We had come to these structures after having visited, earlier in the day, a cave lower on the island in which hunter-gatherers had lived some twelve thousand years prior. And we'd come to the island after having visited a cave on the mainland in which Neanderthals and ancient humans once cohabitated (it was a very good couple of days). In each of these places we'd paused with our two kids to look out over the landscape that these peoples once inhabited. When we did, we also ate. In the Cyclopean landscape, for example, we nibbled on a bit of bread with fresh fig preserves and sipped some of a friend's homemade Plavac Mali wine. In these moments, we wondered what those earlier peoples thought when looking out on the landscapes on which we now looked. It is easy to imagine that some of the things we find beautiful they might also have found beautiful. But we also started to wonder about something else. As we savored our food, we began to wonder what flavors the ancient peoples savored. Did the cyclopean pastoralists, for example, have a favorite cheese? Did the paleolithic hunter-gatherers have favorite berries? How much farther might a Neanderthal go to search out the best-tasting prey? These questions were fun. It was easy, at the end of a wonderful day of exploring, to get lost in them.

Later, we started to read more about the diets of paleolithic and more recent peoples, their diets and their pleasures. As we did we realized that while the diets of the peoples of the past are very often measured and discussed, they are almost never talked about in the way that we would talk about our own meals. Our own meals are, on a good day, about pleasure. Those of the ancients, well, of course, they were about survival. In confronting the past, scientists and other scholars had taken the pleasure and deliciousness out of food.[1]

One of us (Rob) is an ecologist and evolutionary biologist and the other (Monica) is an anthropologist. We imagined that one of our fields must have considered the role of deliciousness in the decisions our ancestors made. But neither had.

Evolutionary biologists talk about the optimal decisions that animals make, without talking about how they make them. Historically, they've often tended to assume animals are a bit like robots, able to measure their environments perfectly and respond. A subset of the scholars who study human hunter-gatherers do the same. Search scholarly papers for "optimal foraging and hunter-gatherer" and you will find hours of reading. But search for the three terms "optimal foraging," "hunter-gatherer," and "flavor" and the pickings are slim and a bit unusual. On the other hand, cultural anthropologists have tended to focus on the unpredictable power of culture. "Culture can make someone ferment a shark or eat ants. Don't try to explain it," or so the literature seemed to suggest. Yet, as we traveled around the world and met with people of diverse cultures, we found that they nearly all talked about food and flavor and what is and is not delicious. This was as true in a thatch house in the Bolivian Amazon as it was in palace in Portugal.

Increasingly, we had the feeling that we had accidentally come upon a radical idea, namely that humans and other animal species prefer to eat delicious things when given the choice. Even as we write this, it is shocking that this idea could be novel, much less radical, and yet it has been ignored. Mostly.

Quite apart from ecology, evolutionary biology, and anthropology exists a field called gastronomy. Gastronomy began with a book called *Physiologie du goût,* published by the French gastronome Jean Anthelme Brillat-Savarin in 1825.[2] Brillat-Savarin worked as a lawyer, a mayor, and later the Councillor of the Supreme Court of Appeal, but history remembers him for his ability to ponder and write about food and eating. The book's title was first translated as "The Physiology of Taste," but it neither dwelled exclusively on physiology, nor did it focus on taste. The English word "taste" now is used to describe the sensations that derive from the taste buds on the tongue. Brillat-Savarin didn't mean taste in this sense. He meant something more like what we would now call flavor, the sum total of the sensory experience of eating including taste, aroma, mouthfeel, and so much more. The book then might more accurately have been titled "The History, Philosophy, and Biology of Flavor and the Pleasure of Eating."[2]

Foods that are a pleasure to eat are delicious; to be delicious is to have exceedingly good flavors, pleasing flavors, sensuous flavors, even voluptuous flavors.[3] At the time that Brillat-Savarin published his book, the study of deliciousness was the territory of bakers, brewers, vintners, cheesemakers, cooks, chefs, gourmands, and gourmets. For philosophers and scientists, the mouth was a backwater, too ordinary and vulgar—all teeth, spit, and tongue—to be taken seriously. Brillat-Savarin took the mouth seriously. Napoleon had been deposed a decade earlier. France was reinventing itself. It was a time for sweeping statements about the world. As a gourmand, Brillat-Savarin would make those statements from the perspective of pleasure in general and deliciousness in particular. He blended what chefs knew, what scientists were beginning to learn, and his own sometimes prescient insights. The book was beautiful and radical. It was also ridiculous and idiosyncratic (including, for example, a list of Brillat-Savarin's favorite sayings such as "A dinner without cheese is like a beautiful woman with one eye"). Despite its quirks, or perhaps in part because of them, the book offered the hypotheses and questions that would ultimately precipitate thousands of discoveries and insights. It was one of the seeds around which gastronomic sciences nucleated.

Books on gastronomy, in the long wake of Brillat-Savarin, considered insights from chemistry, physics, psychology, and, more recently, neurobiology. Richard Stevenson wrote *The*

Psychology of Flavour, a treatise on the meeting of the subconscious, the conscious mind and food.[3] Gordon Shepherd wrote *Neurogastronomy* (which might also have been called "The Neurobiology of Flavor") and later *Neuroenology* (the neurobiology of wine flavor).[4] Charles Spence wrote *Gastrophysics* (the physics of flavor), and Ole Mouritsen and Klavs Styrbæk wrote *Mouthfeel* (a more comprehensive consideration of the physics of flavor).[5] But there was no book that directly considered the evolution of gastronomy or deliciousness in light of human evolution, ecology, and history. We decided to write that book. This, we hope, is that book.

In the pages that follow we build on insights from the fields of human ecology, anthropology, ecology, and evolution, in concert with those of physics, chemistry, neurobiology, and psychology, to make sense of flavor, its evolution, and its consequences. We weave together what chefs now know about the experience of food, what ecologists know about the needs of animals (especially the human animal), and what evolutionary biologists know about how our senses have evolved. In some cases, we develop novel hypotheses, but far more often we simply connect ideas that have not yet been well connected. In doing so, we tell a story of evolution and history that puts pleasure and food where they deserve to be in our drama: at the center. With this book, we hope to enlighten but also to offer practical insights for making more sense of all of the food in your kitchen and why it is (and sometimes isn't) delicious.

Our book is mostly chronological. In chapter 1, we begin with a consideration of the role taste receptors have played over the last several hundred million years in guiding animals toward their needs and away from dangers. We also consider the evolution of differences in taste receptors among species of vertebrates. The hummingbird tastes a different world than the dolphin or the dog. The evolution of taste receptors has guided animals toward their changing needs through deliciousness.

For most of our evolutionary history, our ancestors had little influence on the availability of the foods around them. Yet, once our ancestors began to invent tools, roughly six million years ago, things changed. Our view of this time in our evolutionary prehistory is fuzzy, but modern chimpanzees provide a lens into what it might have been like. Chimpanzees use tools to access food that would otherwise be unavailable; in doing so, they create cuisine. Different chimpanzee communities have different cuisines and, more generally, culinary traditions. But their cuisines are united in their inclusion of foods that are sweeter, more savory, or otherwise more pleasing to eat than what would most easily be available. Sometimes, those foods are integral to survival. Just as often they appear to be relatively unimportant, pleasurable snacks. It seems likely that life was similar for our chimpanzee-like ancestors six million years ago, ancestors for whom flavor and culinary traditions may have played a key role in the advent of the tools that precipitated major evolutionary changes. In chapter 2, we argue that the proximate reason for several major evolutionary changes in our ancestors may have been that they found ways, using tools, to seek out, find, and eat more flavorful foods. The nutrients and energy provided by these foods eventually changed the evolutionary trajectory of our ancestors, but first and foremost this transition was about taste and other components of flavor. In chapter 3, we then discuss the ways in which evolutionary changes in primate heads in general and human heads in particular led the aromas sensed in the mouth (as part of flavor) to play a more important role than they had before.

As our flavor-conscious ancestors invented new tools, evolved bigger brains, and developed more complex cultures, they also began to hunt more. As they did, they began to over-hunt some species. Neanderthals and then *Homo sapiens* in Europe, and *Homo sapiens* in the Americas as well as Australia and nearly each and every island on Earth, contributed to the extinction of hundreds of the largest, most unusual animals on Earth. Five-foot-tall owls disappeared, as did tiny elephants, giant sloths, predatory kangaroos, and many hundreds of other species. A voluminous literature considers just how important human hunting was to these extinctions (the argument is about whether it played the only role, the main role, or a minor role). But essentially no studies consider whether flavor influenced the species our ancestors chose to eat. In chapter 4, we argue, in light of Clovis hunter-gatherers in the Americas, that flavor played a role in the choice of which animals to hunt. Most of the preferred prey species of the Clovis hunters are now extinct, and many appear likely to have been delicious.

One of the consequences of the loss of many of the species ancient hunter-gatherers preferred to eat is that those species are no longer around for us to taste. The feet of mammoths appear to have been particularly delicious, and, well, you will not have the opportunity to try them. But another consequence has to do with, perhaps surprisingly, fruit (chapter 5). Fruits evolved to please animals, but the fruits we most enjoy, many of them anyway, evolved not to please us but instead to please the mouths of species that are now extinct. From fruit, we move on to consider the ways in which flavor aided our ancestors as they began to use spices (chapter 6), then to ferment meat, fruits, and grains (chapter 7). We imagine that our eyes and ears guide us, and yet with regard to both spices and fermentation, we chose with our noses and mouths. It was our noses and mouths that helped to usher in the spice trade and so too our noses and mouths that enabled us to understand how to create (and love) beers, wines, and stinky, fermented fish.

In some moments in history and prehistory, humans chose to create foods that appealed primarily to the sense of taste. In others, they created foods that appealed to taste but also other components of flavor, including mouthfeel, aroma, and more. Such foods include the stinky tofus found across much of Asia, the curries of India, and the washed-rind cheeses of Europe. In chapter 8, we try to understand why, in certain moments, humans choose to make complex labor-intensive foods when other types of foods would be easier (and just as nutritious). We argue that part of the answer is flavor. We do so in the specific context of a group of monks whose work (and pleasures) changed the food of Europe. Finally, in chapter 9 we conclude the book by considering those contexts in which we gather together to feature food, to enjoy food and each other, whether they be around fires or at fêtes. In doing so, we imagine a new future for the study of flavor, one in which everyone is around the table, scientists, chefs, farmers, writers, and shepherds alike, breaking bread or carving stinky tofu, as the case might be.

In short, our human evolutionary story is a story of flavor and deliciousness, and the story of flavor and deliciousness is a story of physics, chemistry, neuroscience, psychology, farming, art, ecology, and evolution. From the telling of the stories of flavor and its evolution and consequences, new insights into our daily food emerge.

In general, the two of us tell these stories together. Over the last twenty years we have shared many of our food experiences and conversations. But sometimes, just Rob was present at a particular meal or event. When that is the case, we refer to him in the third person (Rob . . .). For the most part, though, we've been in it together. We've bored our kids with it (and sometimes interested them—they've both read the whole book).

We've gone to market after market and meeting after meeting and tasted food and drink after food and drink. And so, this book is written by both us, Rob Dunn and Monica Sanchez. Here and there you can hear one of our voices coming out a little more than the other's. (If the text is funny, it's Monica. If it seems like it might be funny but then isn't, it's Rob.)

We did not come to the ideas in this book on our own. When we began to describe elements of flavor, we quickly realized we weren't doing it with the sophistication that a gastronomist such as Brillat-Savarin might. What is more, we also realized, in talking about this book, that part of the great joy of thinking about food in this new way is sharing the ideas, conversation, and food with people with other perspectives. This has been particularly fun on those occasions in which we've had the chance to spend time with people who work with food for a living. Rob collaborated with Anne Madden, an expert on yeast biology, and with a dozen bakers in Belgium, to understand how the lives of bakers influence the flavors of their breads. We both followed a truffle farmer and his dog in the quest for truffles. We went behind the scenes at a distillery in Denmark where we met a brewer who wanted to spend the afternoon talking about the natural history of bees and the ways in which bees employ fermentation. We traveled into a thousand-year-old wine cellar in eastern Hungary to film a documentary, and found ourselves lost in conversations about the fungi growing in the cellar. In these experiences and others, the richness of the conversations has made our thinking clearer, made the food we shared better, and, well, frankly, left us feeling happy and fulfilled.

We've included the names of the many people who helped with this project in the book. In some places, we mention our dinner companions by name in the main text. But where we don't, they are listed as the final endnote of each chapter. These

FIGURE P.1. The tops of some of the "Cyclopean" walls of a pen and, in the background, other ancient structures, on an island in the Dalmatian region of Croatia.

people have been our sounding board. They have chimed in, again and again, to say "Oh, but don't you know, the nuts chimpanzees eat taste like walnuts but with a hint of thyme," or "The smell of dashi is the smell of the seaweed, which is the smell of the sea." Or, sometimes, when our ideas strayed a little too far from what we could actually show with any comprehensiveness, simply, "Bullshit." As a result, this book is more like a dinner party at which we are the hosts than a singular creation of the scientist in the woods or artist in front of clay. The voice in the book is ours, but the ideas have been informed by our companions, companions with whom we are very grateful to have shared the pleasure of ideas and food.

CHAPTER 1

Tongue-Tied

Tell me what you eat, and I will tell you what you are.

Taste seems to have two chief uses: 1. It invites us by pleasure to repair the losses which result from the use of life. 2. It assists us to select from among the substrates offered by nature, those which are alimentary.

—JEAN ANTHELME BRILLAT-SAVARIN,
THE PHYSIOLOGY OF TASTE

The nature of pleasure and displeasure have preoccupied humans since the first paleolithic philosophers sat around a fire, roasting meat and talking. What questions could be more essential than "Why do we experience pleasure or displeasure?" Or, "When and why should we allow ourselves to enjoy pleasure or subject ourselves to displeasure?" In the first century BCE, the Roman poet Lucretius offered an answer. He argued that the world was material, composed of atoms and atoms alone. Atoms made up the moon, the fence, and the cat on the fence. They also made up the mouse upon which the cat was about to pounce. In death, the atoms in the mouse might be rearranged into the body of the cat, but they would continue to exist.[1] In such a world, pleasure was the body's mechanism for fulfilling its material needs. Pleasure led the cat to the mouse. Pleasure was natural; displeasure too. To Lucretius the naturalness of pleasures and displeasures was not a call for hedonism. But it did suggest that a good life could be one in which pleasures were enjoyed and displeasure was avoided. Lucretius recorded his ideas in a moving poem titled *De rerum natura* and typically translated as *On the Nature of Things* or *On the Nature of the Universe*. The poem brought Lucretius's ideas to a large audience. They weren't new ideas, not entirely. In part, Lucretius was reiterating and rewriting the ideas of the Greek philosopher Epicurus. But these ideas were nonetheless given a new clarity and beauty. Yet, when the Western Roman Empire collapsed, Lucretius's words were, bit by bit, lost. By the late Middle Ages, the primary evidence that Lucretius existed was indirect. He could be found in the writings of other scholars, scholars who mentioned and sometimes quoted tantalizingly short excerpts from *On the Nature of the Universe*.

With the fall of the Western Roman Empire, many of the great literary and scholarly works of ancient Romans and Greeks vanished. They were burned, crushed or, more often, simply neglected. Some works were lost permanently. But not all. Many were copied and studied by Muslim scholars in Byzantium; others were preserved in monasteries. Fortunately, Lucretius's poem was among those manuscripts that were saved. In 1417, *On the Nature of the Universe* was found in a German monastery[2] by a restless and curious monk named Poggio Bracciolini.

Poggio was struck by the intense beauty of Lucretius's work. With time, he also became aware that the world Lucretius described, a world full of natural pleasures, seemed to be at odds with everything he had learned as a medieval Christian. He eventually came to criticize the poem, but not before ordering a scribe to make a copy and then sharing that copy around (and having more copies made). In the coming decades some would come to regard the sentiments embodied in Lucretius's poem as a defining model for the future, grounded in the past. Meanwhile, to others Lucretius's ideas were a threat to Western civilization. Our perspectives on pleasure and the materialism of the world remain as divided now as they were then. Such divisions bubble beneath many of our most politicized debates. We won't resolve such debates here, but we can introduce a missing piece, the answer to the question of why pleasure and displeasure exist. Pleasure is caused by a particular mix of chemicals in the brain. So is deliciousness, the specific pleasure associated with the flavors of food. An animal's body produces those chemicals in order to reward it for doing those things that will aid its survival and chances at reproduction. As Lucretius recognized, this is as true for mice or fish as it is for humans.[3] Displeasure is the opposite. It penalizes animals for doing things that make survival and reproduction less likely. Together, pleasure and displeasure are nature's simple way of helping to ensure animals stay alive long enough to make more of themselves and pass on their genes.

One of the things any animal needs is to eat the right food. Just which food a species needs to be guided to, by pleasure, is predicted by a field of science called biological stoichiometry. *Biological stoichiometry* is perhaps the most boring possible name for a field with enormous consequences for how the world works. It is an obscure field. If you don't study biological stoichiometry, you have probably never heard of biological stoichiometry.

Biological stoichiometry concerns itself with balancing various versions of a single equation. In the simplest version, the left side of that equation is made up of the bodies of organisms that have been eaten (the prey). Think about all of the animals, plants, fungi, and bacteria you have consumed in your own life. The right side of the equation is the body of the organism doing the eating (the predator), along with all of the waste it has ever produced and all of the energy it has ever used. As Lucretius put it, animals "borrow lives from each other."[4] They are relay runners that "pass along the torch of life." Biological stoichiometry deals with the rule by which the baton is passed.

Stoichiometry's rule is that the equation must balance; the nutrients present in the food and those in the consumer (and its waste and consumed energy) must ultimately match. This is where things get trickier, where the problem begins to resemble an elementary school homework question with a man and two dogs on one side of the river and a woman and a canoe on the other. If the body of a predator, for example, has a high concentration of nitrogen, so too must its prey. This seems so obvious as to not even bear writing down. Brillat-Savarin told us this: you are what you eat and you need to eat what you are. But the tricky part is that the equation linking predator and prey relates not just to, say, nitrogen and carbon; it also relates to any other nutrients that the predator cannot make for itself. As a result, the predator and prey must balance with regard to nitrogen but also magnesium, potassium, phosphorus, and calcium, each of which plays a role inside every animal cell.

We can actually write out the proportional number of molecules of each element present in the bodies of different species of animals (and hence the predator, or more generally, consumer, side of the equation). The average mammal, for example, can be described chemically by the list of elements in its body and their relative proportions. Here is the ingredient list for making a mammal:

$$H_{375,000,000}, O_{132,000,000}, C_{85,700,000}, N_{64,300,000}, Ca_{1,500,000},$$
$$P_{1,020,000}, S_{206,000}, Na_{183,000}, K_{177,000}, Cl_{127,000}, Mg_{40,000}, Si_{38,600},$$
$$Fe_{2,680}, Zn_{2,110}, Cu_{76}, I_{14}, Mn_{13}, F_{13}, Cr_7, Se_4, Mo_3, Co_1$$

Mammals, such as humans, have 375,000,000 times more hydrogen (H) atoms in their bodies than cobalt (Co) atoms. Today, scientists can calculate the elemental ingredient lists of humans and other mammals with great precision. But how do wild mammals know how to find all of these elements in nature in order to have what their bodies need and balance their own stoichiometric equations, equations in which the ingredients they consume match those their bodies need?[5] How does any animal know? How, for that matter, do you know?

For predators that eat their prey's muscles, organs, and bones, hunger (and the pleasure triggered when hunger is sated) might be enough to balance the equation. Dolphins need only hunger and some kind of mental image of what food looks like when compared to non-food (something that tells them not to eat a rock).[6] Things are mostly in balance.

For animals with diets that allow them more choices, things get trickier. For animals that eat plants (herbivores) or animals and plants (omnivores) life is especially challenging. As can be seen in figure 1.1, many elements are found in far higher concentrations in animals than in plants. If an omnivore randomly eats some plants and some animals, it will easily end up with a diet that is deficient in sodium, phosphorus, nitrogen, and calcium. Things are just as tricky for herbivores. How do herbivores and omnivores know how to balance their own stoichiometric equations? To a large extent, they make decisions based on flavor. Flavor is the summation of all of the sensory experiences that occur inside an animal mouth. Flavor includes aroma, mouthfeel, and also taste.[6] Each of these components of flavor is important in guiding animals toward their needs, but taste plays a special role.

The English word *taste* comes from the vulgar Latin *tastare*, which some dictionaries contend is an alteration of the Latin word *taxtare*, "to handle or grasp." This alteration may be due to the influence of the Latin word *gustāre*, which means to taste. When we taste, we grasp with our tongues. The tongue is covered in taste papillae (the bumps you see in the mirror) in which are found taste buds each of which contains taste receptor cells layered like petals within a flower.[7] These cells are replaced every nine to fifteen days. Even as a vertebrate animal ages, its tongue is always being reborn. Tentacular hairs project from each taste cell. At the tip of these hairs one finds the actual taste receptors, waving in the mouth's tumultuous sea.

Each type of receptor is a like a lock that can be opened only by a specific key. Open the lock with the right key and a signal is sent from the taste receptor along nearby neurons. From there, the signal splits and travels via separate nerves to each of several parts brain. One of the signal's paths reaches the primitive, ancient fish part of the brain that controls breathing, heart rate, and other subconscious, necessary, elements of the body's working. For tastes associated with elements that are needed—such as salt or sugar—one effect of the signal's arrival in this primitive part of the brain is the release of dopamine. Dopamine triggers a flush of endorphins which you experience as a vaguely conscious sensation of pleasure; it is a pleasure that rewards animals for finding what they need. It also creates cravings: "I love this, I want more." Another of the signal's paths reaches the conscious part of the brain, the cortex. Once there, it triggers the specific sensation associated with what has been tasted, such as "salt," or "sugar."[8]

This taste system works because the elements any particular animal needs are relatively predictable. They are predictable based on the past: what an animal's ancestors needed is likely to be what that animal also needs. Taste preferences, therefore, can be hardwired. Consider sodium (Na). The bodies of terrestrial vertebrates, including those of mammals, tend to have a concentration of sodium nearly fifty times that of the primary producers on land, plants (figure 1.1). This is, in part, because vertebrates evolved in the sea and so evolved cells dependent upon the ingredients that were common in the sea, including sodium. To remedy the difference between their needs for sodium and that available in plants, herbivores can eat fifty times more plant material than they otherwise need (and excrete the excess). Or they can seek out other sources of sodium. The salt taste receptor rewards animals for doing the latter, seeking out salt in order to reconcile their great need and balance their life's stoichiometric equation.

Most mammals appear to have two kinds of receptors that respond to the sodium (Na) in salt (NaCl). One of the taste receptors responds to sodium above a certain minimum threshold concentration. If sodium is present above that concentration, it sends a signal to the brain. Pleasure ensues, as does the conscious perception of "salt." Think of biting into a big soft *laugenbrezel* at the little shop between the airport and the train station in Berlin (or at least that is what we thought of while writing this). This first receptor leads mammals toward salt. For example, elephants walk hundreds of miles to muddy patches of salty soil. In doing so, they wear game trails deep into the ground, trails that trace the geography of their needs.

But as much as not eating enough salt (and hence sodium) is bad, eating too much salt can also be bad. The ingestion of too much salt can easily occur in mammals that live by the sea if they slake their thirst with salt water. To cope with this potential problem, mammals have a second salt taste receptor that detects high concentrations of sodium and, having done so, sends a signal of displeasure and a conscious perception of "too much!" to the brain. If you get a particularly salty bite of your laugenbrezel and feel compelled to brush off some of the salt, it is this second receptor at work. Salt taste receptors lead terrestrial mammals, whether they be mice, squirrels, or humans, toward the concentrations of salt that, on average, they and other terrestrial vertebrates have tended to need over the last tens of millions of years. They lead them toward those concentrations and, simultaneously, away from excess.

Lucretius imagined that fatty foods might be made up of smooth atoms and bitter or sour foods crooked, rough, and barbed ones. They aren't. Instead, the experience any animal has of a particular food reflects how its taste receptors are connected to its brains. The sensation we experience associated with salt, the sense of the taste "salt," is entirely arbitrary. We can know that other animals have salt taste receptors just like our own and we can know that those receptors trigger cravings and pleasure (thanks to detailed studies in mice and rats) and even at what concentrations, but we cannot know what "salt" *tastes* like in those other species. We don't know exactly what the pleasure of encountering such a taste *feels* like in those other species. We don't know anything about the experiences of tastes or pleasures in humans other than ourselves. We just assume they are always the same.

As you can see in figure 1.1, sodium (Na) isn't the only element that is more common in vertebrate bodies, such as those of mammals, than in plants. So too is nitrogen (N). In plants

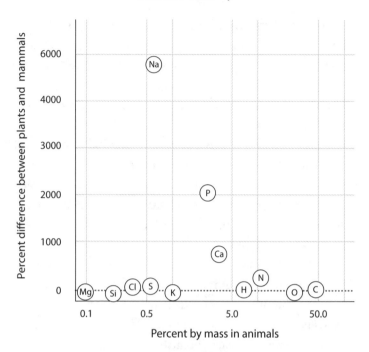

FIGURE 1.1. The percent by mass of the elements most abundant and biologically "essential" in animals (horizontal axis), and how these compare with their abundance in plants (vertical axis). Elements with a positive values are more concentrated in animal than plant tissues. For example, sodium (Na) is nearly 50 times (or 5000 percent) more concentrated in the bodies of mammals than in the tissue of plants. Conversely, silica (Si) is slightly more concentrated in plants than in animals.

and animals, nitrogen tends to be found in the amino acids and in nucleotides. Amino acids are the Lego bricks out of which proteins are made, and nucleotides are the bricks from which DNA and RNA are built.

Animals that eat some plants, be they pigs, humans, or bears, can easily end up with diets deficient in nitrogen. On average, animals have about two times as much nitrogen as plants, as a proportion of their body mass. How do omnivores and herbivores deal with this shortage? Some species just consume two (or more) times as much food as they need and void the excess. Like aphids, scale insects, for example, drink from the sugary phloem flowing through plant veins. In doing so, they gather the small amounts of nitrogen in what they have imbibed and as much sugar as they need, then excrete sugar water. That excess falls from them and is gathered by ants and some humans as a delicacy. (It is thought the manna of the Bible may have been the excess falling from the tamarisk manna scale insects, *Trabutina mannipara*, feeding on tamarisk trees.) But for mammals, this approach to balancing isn't a great solution. Instead, a taste receptor for nitrogen, or one or another compound that is indicative of foods with nitrogen, seems like a better approach. But until 1907 no taste in humans was known to correspond to the presence of nitrogen, or the amino acids and proteins in which nitrogen is found, in food.

In 1907 Kikunae Ikeda, a chemistry professor at Tokyo Imperial University, was eating a bowl of broth that changed his life. The broth was dashi. Ikeda had consumed dashi before, but on this particular occasion he was struck by its deliciousness. It was salty, a tiny bit sweet and, well, there was a hint of something else, something very good. Ikeda decided he wanted to identify the origin of this extra taste, the very good taste that he would come to call "umami." The word "umami" is rooted in the Japanese words for savory (*umai*) and essence (*mi*). It also means "a delicious taste and its level of deliciousness," as well as "a skillful thing to relish, especially in relation to techniques in art."

The recipe for dashi is superficially simple. It includes fermented fish flakes (katsuobushi),[9] water, and, in some cases, a special kelp (kombu). Ikeda knew the taste did not come from the water. It must then have come from either the fish flakes or the kombu. All Ikeda had to do was identify which compound in the fish flakes or kombu triggered the taste he believed himself to have perceived, the taste of umami. This was easier said than done. A "simple" dashi broth can contain thousands of chemical compounds potentially able to produce tastes or aromas. Ikeda had to identify these compounds and test them one by one. According to the tally of Jonathan Silvertown in his book *Dinner with Darwin*,[7] it took thirty-eight separate steps to finally extract some gritty crystals from the kombu kelp in the broth that appeared both to be relatively pure (a single compound) and to taste of umami. The crystals were glutamic acid. Glutamic acid is an amino acid; it is a building block of protein and so a reliable indicator of the presence of nitrogen in a food. The taste of umami is a taste that rewards us for finding nitrogen. Umami taste, triggered by glutamic acid, leads us toward our necessary amino acids. But umami taste is not triggered by glutamic acid alone.

Subsequent studies by other Japanese researchers would show that in addition to glutamic acid, inosinate and guanylate, two ribonucleotides, can also trigger umami taste. These two ribonucleotides are not found in the dashi's kombu, but instead in the fish flakes. When inosinate or guanylate and glutamic acid are experienced together, they produce a kind of super umami. Glutamic acid and inosinate are experienced together in dashi. Dashi is rich with super umami, a flavor that is both deeply pleasing and indicative of the presence of nitrogen.

For decades, few scientists outside of Japan believed Ikeda's result (nor, for that matter, the subsequent results related to inosinate and guanylate). But don't feel too bad for Ikeda; he patented the method used to produce MSG in 1908. MSG results from the combination of glutamic acid and sodium. Thanks to that patent, Ikeda did just fine for himself.[8] People were willing to pay for umami taste even before they believed it to exist. As for why Ikeda's work was neglected outside of Japan, it was partly because the first paper was written in Japanese and so not widely read by scientists in Europe and the United States. But it wasn't just language, it was also a problem of mechanism. Although Ikeda could show that when his glutamic acid crystals were added to a food that they improved its taste, he hadn't identified how the mouth tasted them. The taste receptor for umami would not be discovered for ninety years. The separate receptor that responds to inosinate and guanylate would take even longer to resolve. It was only once they were discovered that umami taste was widely accepted by most sensory scientists as a human taste.

If you return to figure 1.1, you will see that another element that is much more common in animals than in plants is phosphorus (P). Phosphorus is more than twenty times as concentrated in the bodies of animals as in the tissues of plants. A lack of phosphorus is a key challenge faced by many animal species. [9] Why then isn't there a taste receptor that detects phosphorus in food and rewards animals for finding it? One possibility is that foods, particularly foods in the form of whole animals with lots of nitrogen, typically also have sufficient phosphorus. Perhaps having a receptor for one of the two nutrients was sufficient. Nature often packages nitrogen and phosphorus together. Yet, this wouldn't explain how herbivores or even most omnivores find phosphorus. Another possibility is that some animals do have a phosphorus taste receptor.

Michael Tordoff is a scientist at the Monell Chemical Senses Center (in the world of taste, all roads lead to Monell). He has specialized in laboratory studies of poorly charted tastes, including the taste of phosphorus. Since the 1970s, studies have shown that mice are able to somehow taste phosphorus salts. More recently, Tordoff was able to show that mice appear to be able to distinguish between low concentrations of such salts (which please them) and high concentrations (which displease them).[10] Tordoff suspects that most mammals, including humans, have the ability to taste phosphorus salts and to distinguish pleasing concentrations of such salts from displeasing ones.[11] With the discovery of umami, the broad acceptance that umami was a taste required the discovery of the taste receptor for umami and its functioning. Tordoff is on his way to that step with phosphorus. Recently he even discovered what appears to be the receptor that alerts mice that they have found too high of a concentration of the phosphorus (in the form of phosphates).[12] No one has yet discovered the receptor that tells them when they have found just the right amount. It is possible that someday soon phosphorus may be accepted as an additional human taste.

You might imagine that the discovery of a new taste, a taste that you might be experiencing each time you eat, would trigger hundreds of follow-up studies. An award of some sort. Television interviews. It hasn't yet. The world is full of mysteries. Even mouths are full of mysteries. As a result, Tordoff's studies of the taste of phosphorus are cited by relatively few other papers. One of those papers demonstrates that cats, like mice, prefer foods that contain more phosphorus. Phosphorus is now added (as phosphate) to most cat foods to encourage cats to eat the food. Cats don't need to believe or not believe Tordoff's results in order to experience the pleasures, it seems, of phosphorus taste. Meanwhile, the other element that is scarce in animal diets relative to animal bodies is calcium. Tordoff thinks he has discovered evidence of a calcium taste receptor too.

Most of the elements and compounds we need in our diets are necessary for building new cells and other components of our bodies. Because of this, we need them in proportion to their relative rarity or abundance within our bodies (that equation again). In addition, however, our bodies also need energy for daily activity; even once the building is built you have to keep the lights on. The more active a species is, the more such energy it needs. This is as true for insects as it is for mammals. The most active, aggressive, ants, for example, require the highest calorie diets.[13] Most of that caloric energy, whether for ant or elephant, comes from breaking apart carbon compounds.

Simple sugars, all of which are small carbon compounds, are easy for animals to convert into energy. Simple sugars include glucose, fructose, and the result of their biochemical marriage, sucrose. Sweet taste receptors reward animals for finding these sugars.[10] They reward us with sweetness for eating mangos, honey, figs, or nectar. Complex carbohydrates, such as starches, are also sweet to many mammals. Old world monkeys, apes, and humans are unusual in that their sweet taste receptors do not respond to starch. However, these species produce an enzyme called amylase in their mouths. This amylase does not aid in the digestion of starch (which happens later) but has been hypothesized to break down some of the starch in the mouth so that it can be detected by the sweet taste receptor. Ancient humans, like modern gorillas or chimpanzees, produced some amylase in their mouths but not much. However, with shifts to more starchy diets, some groups of humans evolved the ability to produce more amylase in their mouths, perhaps to more quickly perceive starch to be sweet. Evolution can make bland foods sweet and vice versa, simply by changing how they are perceived.

The other source of energy for working cells is fat (protein can also be converted to energy, but is the body's last choice). Fats contain twice as much energy per gram as do simple sugars. Not surprisingly, many mammals appear to experience pleasure in eating fat. For example, Danielle Reed (yet another scientist at the Monell Chemical Senses Center) used to give her laboratory mice a high fat diet. When she did they would, as she put it, go on a "Friday night binge. They would just eat all their fat and groom their hair with it and they'd just get in the middle of their fat. They love fat."[11] Surprisingly, it is not clear what it is about fat that mice or other animals enjoy. The answer may be mouthfeel. Fats have a pleasing mouthfeel (a gastronomic term for the sensation of touch as it is manifest inside the mouth). Put a piece of avocado in your mouth. It will be pleasing, but the pleasure is not the taste (it is not very sweet, nor sour, nor salty, nor really umami). Nor is the pleasure of the avocado its aroma, which is simple, often described simply as "green." The pleasure is, instead, the feel, the smooth touch of the fruit, the same smoothness we experience when enjoying butter or cream. This touch is part of the story.[12] But mysteries remain.

Salty, umami, and sweet taste receptors (and maybe also phosphorus and calcium taste receptors) evolved to point animals, through deliciousness, to what might otherwise be missing from their diet, whether in order to make new cells or, in the special case of simple sugars, to make new cells and to run them. But taste receptors can also serve the opposite purpose; they can point animals away from danger. They do so through feelings of displeasure. In some contexts, sour taste, which detects acidity in food, is displeasing. We will return to why this might be in chapter 7 (sour taste is mysterious and yet potentially very important to our human story). The more clear-cut case is that of bitter taste receptors. Bitter taste receptors allow animals to identify plants, animals, fungi, and anything else in nature that might be dangerous to ingest. For nearly all taste receptor types, animals have one or two (salt) basic classes of receptors. With bitter taste receptors, animals have many kinds.

Each kind of bitter taste receptor is triggered by one or more chemicals or classes of chemicals. Lucretius wrote of "nauseous wormwood," a key ingredient in absinthe, whose "foul flavor set the lips awry." We now know that it is the absinthin in wormwood that triggers one of our bitter taste receptors. And we even know which receptor (hTAS2R46, if you are curious). A different receptor responds to strychnine in plants; another responds to the noscapine found in poppies and their relatives. Yet another responds to the salicin in willow bark (and aspirin). Because being able to avoid toxic chemicals is very important (and failing to do so often results in having no offspring and so not passing on your genes) bitter taste receptors tend to evolve relatively rapidly. Species tend to have bitter taste receptors that reflect the dangerous kinds of compounds they are most likely to find in their environments. Humans and mice, for example, have about 25 and 33 kinds of bitter taste receptors, respectively, but the overlap between ours and theirs is modest.[14] Some compounds that mice evolved to avoid (and hence taste as bitter) have no taste in our mouths and vice versa. Variation even exists among humans within populations. As Lucretius put it, "what is sweet to some, to others proves bitter." As a result, a group of people might be able to detect more kinds of compounds as bitter than any individual. The combined knowledge of a community contains three types of bitter compounds then, those that everyone tastes as bitter (dangerous), those that some people think are bitter (maybe dangerous) and those that no one tastes as bitter (safe).

But, although most vertebrate species can detect many kinds of potentially toxic compounds via many types of taste receptor, and different individuals are able to taste different compounds as bitter, individual vertebrates perceive only one kind of bitter. All the bitter taste receptors are wired to a single nerve and only register a single conscious perception BITTER.[13] If a bitter compound is ingested in a high concentration, it can trigger nausea. If it is ingested at a high concentration twice (for example, via two gulps) the stomach muscles of the consumer stop contracting in rhythm. They begin to twitch out of sync which ultimately, if the dance of indigestion is sufficiently strong, triggers vomiting. Bitter taste receptors tell us things are bad and then, with vomiting, trigger both a reminder that they were serious and, with that reminder, expel some of the offending compound.

The displeasing sensation a species experiences in association with bitter compounds is just as arbitrary as that of saltiness or sweetness. Its key message is simply displeasure, displeasure that, like a stick, is meant to lead animals from things they are too stupid to avoid otherwise.[14] As humans we have learned to sometimes ignore the bitter taste warning these receptors offer us, such as when we drink coffee, hoppy beers, or bitter melons. We do so even as our tongues cry out, "Bitter. Danger. Bitter. Danger." "Hush now," we say to our tongues as we enjoy coffee, tea, or hoppy beer. "Hush, I know how much of this toxin I can consume without danger. Hush, I know what I am doing. I have learned."

What we've just described of the taste system is representative of the average terrestrial vertebrate. But as terrestrial vertebrates have evolved, their lifestyles have changed. Such changes have led to (or in some cases been caused by) evolutionary changes in taste receptors, such that each species perceives, with its mouth, a different world. Or, as Lucretius put it, "there are different senses in living creatures, each of which perceives in itself the object proper to it."[15] Some of the changes are subtle and relate to the thresholds at which particular compounds are detected. Others of the changes are more extreme and include the losses of entire tastes.

TABLE 1.1. Taste Thresholds for Humans

Taste	Substance	Necessary concentration to trigger response (parts per million)
Salty	Sodium chloride (NaCl)	2000 ppm
Sweet	Sucrose	5000 ppm
Umami	Glutamate	200 ppm
Sour	Citric acid	40 ppm
Bitter	Quinine	2 ppm

The minimum concentration of a substance needed to trigger a taste receptor varies greatly from receptor to receptor. Bitter taste receptors tend to be triggered by even very low concentrations of the chemicals to which they respond, such as quinine, a toxin produced by plants. These receptors evolved in order to warn us away, and that works best if it happens before we ingest a lot of whatever it is that has touched our tongue. Sugar, on the other hand, is most useful if it is in high concentrations. Below such concentrations our tongues don't even know they've encountered something sweet. The other taste receptors fall in between. Sour is the most unusual of the taste receptors. It deserves special treatment so we'll return to it in chapter 7. The data shown here are for a subset of studied humans. These thresholds, however, differ among species as well as among individual humans.

Perhaps the fastest of the slow ways taste receptors evolve is by breaking. Taste receptor genes tend to be large and so are prone to collect mutations that break them so they can no longer function. Over millions of years the genes for particular taste receptors have broken again and again when the desires (or avoidances) of an animal and its needs are mismatched. Cats, be they pumas, jaguars, or house cats, are strict carnivores (though see, in chapter 4, the special case of cats and avocados). Cats have evolved specialized forms of hunting so as to be extraordinarily good at killing their prey. If you look again at figure 1.1, you will see that an animal that only eats other animals will tend to have in its diet about the right concentration of nitrogen and phosphorus. It also ends up with enough energy, in the form of fat and sugars in its prey's cells, to carry out its daily activity. Cats with sweet taste receptors are no more likely than those without to survive and flourish; if they spent too much time sipping nectar and too little time eating prey they might have even been less likely to survive. As a result, when the sweet taste receptor of an ancient cat broke, that cat survived nonetheless. It did more than survive, as Xia Li (at the time also a researcher at the Monell Chemical Senses Center) recently showed. It begat all modern species of cats. No modern cat species have functioning sweet taste receptors.[15] Forests of sweet fruits and nectar are not delicious to cats, not even a little. If you give a cat a sugar cookie, well, it really doesn't care. It does not experience any pleasure in the cookie's sweetness; the cookie, to the cat, is not sweet.

Like cats, other carnivores such as fur seals, Asian small-clawed otters, spotted hyenas, fossa, and bottlenose dolphins also have broken sweet taste receptors. All of these breaks in the sweet taste receptor gene occurred independent of each other; they are convergent forms of falling apart. One question one might ask about these carnivores is why others of their taste receptors haven't also broken. Cats are unlikely to need more salt than their prey contain. That the cats' salt taste receptors, as well as those of other carnivores, haven't also broken may just be a matter of time. Sea lions have broken sweet taste receptors and broken umami taste receptors. Dolphins have taken this trend further. They no longer taste sweet, salty, or even umami.[16] They thrive on the basis of hunger and satisfaction alone, hunger, satisfaction, and the belief that anything that moves like a fish is dinner. This raises the question of just what it is about a prey item that pleases a dolphin. We don't know. The pleasures of dolphins, whatever they are, are beyond the understanding of science, at least for now.

The loss of particular taste receptors is not the unique purview of predators. Losses have also occurred in animals with diets that are specialized in other ways. The ancestors of giant

FIGURE 1.2. Giant panda surrounded by its one true delicacy.

pandas were bears. As bears, they were omnivores, drawn to living prey but also sweet berries and sour ants. But giant pandas evolved to take advantage of a new diet, one dependent on bamboo. On bamboo alone, they thrive. Initially, as they shifted to bamboo they enjoyed both the bamboo and meat. But with time, giant pandas that were still drawn to meat were either no more likely to survive and mate, or, even less likely, their wants and needs mismatched, their attention distracted. With time the umami taste receptors of giant pandas, like the sweet taste receptors of cats, broke.[17] Now, even if offered meat, giant pandas decline.[18]16

It is unlikely that the descendants of cats, sea lions, or dolphins will enjoy sweetness even long into the future, nor will giant pandas enjoy savory tastes, even though their preference for bamboo has led their populations to decline, in lock step, with declines in size of bamboo forests.[19] It is harder to make something from scratch when it is needed than to break it, a lesson from evolution for daily life. Harder, but not impossible.

Sweet taste receptors, for example, have been lost, but they have also been regained. The ancestor of all modern birds, mammals, and reptiles lived about three hundred million years ago. That ancestor appears to have been able to taste salty foods, savory foods, and sweet foods. However, the ancestor of all modern birds lost its sweet taste receptor. For reasons that cannot yet be discerned, the sweet taste receptor was no longer useful. As a result, birds cannot detect sweetness. Or at least most birds can't.

Hummingbirds descend from ancient swifts. Like modern swifts, these ancient swifts were exclusively insect eating. The ancient swifts were pleased by umami tastes, such as those associated with the bodies of insects or worms, but disinterested in sugars. However, roughly forty million years ago, one population of swifts began to feed on nectar and other sugar sources, perhaps simply to slake their thirst. The nectar was not sweet to the birds. To the extent to which it tasted like anything, it tasted like water. But unlike water, the nectar provided sugars. It has been hypothesized that individuals that drank more nectar were more likely to get energy and pass along their genes, so much so that their umami taste receptor evolved so as to be able to detect sugars in addition to the compounds that ordinarily trigger umami taste (amino acids such as glutamic acid as well as some nucleotides). This swift lineage would become the first hummingbird. Hummingbirds, unlike most birds, can taste sugars and amino acids. However, because they do so using a single receptor it is likely that they experience the two substances as the same, pleasurable sensation, sweet-umami.[20]

These examples of the ways in which an animal species can come to find new things delicious and, in doing so, remedy its deficiencies, are beautiful. They are the fine tuning of the ability of organisms to satisfy their needs through pleasure. The more we study the evolution of taste receptors, the more these stories seem to emerge. We can even predict where they might occur. Hummingbirds are not the only birds that feed on nectar. Sunbirds, flower-piercers, and honey-eaters are unrelated to hummingbirds, but they also feed on nectar and other sweet foods. It seems likely that they too have evolved the ability to detect sugary foods and be pleased by them. Three different desert mammals, in different deserts, have evolved the ability to feed primarily on plants that exude salt. Doing so required them to evolve extraordinary traits that make this lifestyle possible, such as hairs in their mouths that help to scrape salt from the plants. These salty-plant-eating mammals have no need to seek out extra salt and so it seems likely that they have lost their salt taste receptors.[21] But all this fine tuning raises an interesting question when we consider our own lineage.

We are primates, which is to say we are related to lemurs, monkeys, and apes. Within the primates, our narrower branch is that of the hominids, which includes us as well as gorillas, chimpanzees, bonobos, orangutans, and an entire zoo of extinct relatives. Within the hominids, we are the sole surviving member of the tribe Hominini, the hominins. If we look across the entirety of the primates, species differ greatly in their taste receptors. They differ both in what their receptors detect and the thresholds at which they detect them. Some plants that are bitter to us (and deadly) are not bitter (nor dangerous) to some of the monkeys, for instance. Additionally, while we appreciate foods with a relatively low concentration of sugar to be sweet, marmosets only perceive foods to be sweet if the sugars are highly concentrated. In other words, comparing species across the entirety of the primates we see differences, some of them quite big. But then here is the curious thing. If we compare ourselves to our closest living relatives, the chimpanzees, our taste receptors are actually very similar to their taste receptors. What is delicious to a human is, for the most part, delicious to a chimpanzee. This is surprising since, in the time since our shared ancestor, we and chimpanzees have embarked on radically different culinary paths. Chimpanzees live in the forest and, to a lesser extent, grasslands, and eat fruit, insects, and the occasional leg of monkey. We colonized nearly all of terrestrial Earth. As we did, we came to eat something different in each new habitat. Why hasn't the difference between our diet and that of chimpanzees precipitated some kind of major change in taste receptors? In part, the answer is that there have been some subtle changes, if we look closely enough. But there is something else.

When our ancestors began to develop culinary traditions and tools, they found ways to take the foods of any habitat and alter them so as to make them more delicious. In doing so, they dulled natural selection's effects on their taste receptor genes. They dulled nature's effects on which versions of such genes were passed one generation to the next. Our ancestors did not have to wait for natural selection to solve dietary deficiencies through the differential survival and reproduction of individuals with more locally relevant taste receptor genes. They compensated for bland diets by using tools to seek out flavor. Those flavors were often (though not always) indicators of what they needed. This is what Lucretius might have called a "swerve." Through a modicum of consciousness and a pinch of free will our ancestors altered their situation. In doing so they changed the world. In seeking deliciousness, they caused a swerve in the story of their kind, of our kind. This swerve, as we'll argue in the next chapter, was a key step in the evolution of our ancestors.

They figured out how to make tools to find foods that were tastier than those that were otherwise available. They used tools to make their habitats more delicious, then they used tools to help make the landscapes wherever they traveled more delicious. In this way, the pleasure of deliciousness was central to human evolution.[17]

CHAPTER 2

The Flavor-Seekers

*Man alone can dress a good dish; and every man whatever is more
or less a cook, in seasoning what he himself eats.*

—JAMES BOSWELL, *JOURNAL OF A TOUR TO THE
HEBRIDES WITH SAMUEL JOHNSON*

You are not a chimpanzee. Your ancestors and those of chimpanzees diverged roughly six million years ago. The ancestors
of chimpanzees continued to evolve and change in those subsequent years, just as did our own ancestors. Yet, chimpanzees
appear to live lives similar in many ways to those of our shared
great, great, great . . . grandparent.[22] As a result, we can understand aspects of our own past, including the flavors of our past,
by studying the lives of modern chimpanzees. This idea isn't
new. Charles Darwin said as much in 1871 in his book *The Descent of Man*. But it wasn't well appreciated until the early 1960s,
when Jane Goodall began to live with and study the chimpanzees in a seasonally dry forest called Gombe in Tanzania.

At the time that Goodall began her work, chimpanzees were
thought of by scientists as our closest relatives and yet,

simultaneously, not so very different from other primates, be they gorillas or monkeys. Chimpanzees weren't yet a lens into our ancestry. They were just another primate eating fruits in the forest. But the clues were there.

Some of the clues related to the use of tools by chimpanzees. In 1888 Darwin wrote, "it has often been said that no animal uses any tool; but the chimpanzee in a state of nature cracks a native fruit, somewhat like a walnut, with a stone."[23] In the years after Darwin, travelers commented upon the ways in which chimpanzees seemed to use stones to smash nuts. One chimpanzee individual was observed poking a stick into a ground bee hive and licking off the honey. But these observations tended to be described in ways that were dismissive of the abilities of the chimpanzees, as though in each instance the individual chimpanzees had accidentally stumbled upon a rudimentary trick. Give a chimpanzee a typewriter and enough time and it can write the *Odyssey* or, the suggestion seemed to be, manage to use a stick. Yet, as Goodall began to habituate and watch the chimpanzees at Gombe, our collective understanding quickly changed.

Goodall paid particular attention to what the chimpanzees ate and how. She soon saw them using tools repeatedly. They used sticks to probe termite mounds.[24] Goodall observed the chimpanzees making and using sticks to gather termites ninety-one times in 1964 alone. They also used sticks to gather ants. In gathering ants, the chimpanzees of Gombe broke off a relatively consistent length of stick. They then poked the stick into the nest of either army ants (a *Dorylus* species) or tree-dwelling acrobat ants (a *Crematogaster* species), whereupon the ants attacked the stick and could be retrieved and eaten with one pass of a chimpanzee's big, opposable lips. The tools were being used by the chimpanzees as a kind of wooden kitchen utensil—if not quite a butter knife, at least something belonging in the same drawer. Christophe Boesch, who began to study chimpanzees in Taï Forest in Côte d'Ivoire about fifteen years after Goodall began her work in Gombe, thinks of these utensils as being akin to chopsticks; like chopsticks, the sticks could be used in different contexts to different ends. Like chopsticks, they come in different (and yet related) forms.

As Goodall, Boesch, and other researchers studied chimpanzees in more detail, they found that the use of tools was extraordinarily varied. With time, chimpanzees would be observed using leaf tools to scoop up water, stick tools to gather and eat ants, honey (and with it, bees), and algae,[25] and rock tools to smash, open, and eat hard to crack nuts.[1] The more chimpanzees are studied, the more new kinds of tool use are documented. Just as significantly, the tools chimpanzees use and what they use them for differ from one study site to the next. A few types of tool use seem to be restricted to just one place by a single chimpanzee community. In a community of savannah chimpanzees in the Fongoli area of southeast Senegal, for example, female and juvenile chimpanzees hone pointed spears out of sticks with their teeth. They then thrust those sticks into the holes in trees in which fuzzy, big-eyed bush babies sleep, to impale them as giant kebabs.

The differences in tool use among chimpanzee communities are not just due to differences in the habitats they occupy. Two communities in the same basic habitat might use tools for very different reasons. For example, as we've already noted, the chimpanzees at Gombe use tools to access termites. They also use them to gather and eat species of acrobat ants and army ants. At Mahale, another long-term research site just 140 kilometers to the south of Gombe, chimpanzees also eat ants, but they never eat acrobat ants or army ants even though both are found at Mahale. Instead, they eat carpenter ants (*Camponotus* species).[26] These two chimpanzee communities use tools to extract different menus of items from similar surrounding habitats. Different chimpanzee communities also often use different tools to eat the same food, or use the same tool in different ways to eat the same food.[27] These differences in what different chimpanzee populations choose to eat and how reflect the chimpanzees' culinary traditions.

In humans, cultural differences in diet have many components. They include which species a particular group of people eats, how they procure those species, how they prepare those species, and even how they think and talk about those species. Among those who study chimpanzees, however, culture and cultural differences are defined more narrowly. For those who study chimpanzees, the term "culture" is applied specifically to those cases in which different chimpanzee populations use different tools to access the same food or the same tools to access different foods. To avoid entering into the debate about what is and is not "culture," we use the term "culinary traditions" to encompass the range of aspects of diet that differ from one population to the next across generations. Some aspects of culinary traditions might require young animals to watch and copy adults, but other aspects are less conscious. Preferences for the aromas associated with certain flavors, for instance, can be passed by mothers to their babies in utero without any teaching (we discuss such preferences in chapter 6).

Cuisine is defined as "any socially practiced, purposeful transformation of food outside the body.[28] The culinary traditions of chimpanzees, with their tools for turning the inaccessible bits of nature—be they ants, termites, or the innards of bush babies—into food, represent wild cuisines. Cuisine is a major evolutionary innovation. It is rare in nature. Among extant species it is nowhere as elaborated as it is among

FIGURE 2.1. Chimpanzee pounding a nut with a stone hammer.

chimpanzees and humans. Among extinct species, it seems likely to have also been, as we now consider, a key feature of the lives of our common ancestors. Its origin occurred in light of flavor and deliciousness. We imagine the common ancestor of humans and chimpanzees six million years ago as a species living, like chimpanzees, in multiple communities, each with its own culinary traditions.[29] Eventually at least some communities of this ancestral ape figured out how to use the many kinds of utensils modern chimpanzees use. They would have used rocks to smash open nuts. They would have had diverse sticks for collecting diverse kinds of honey and specialized sticks, or even sets of sticks, to gather termites and ants. And they would have had tools for hunting other mammals, sometimes, but probably not so very often. This was not yet the Stone Age. It was instead the Stick Age, the age at the very beginning of culinary traditions and cuisines.[2]

During the Stick Age the climates of Africa began to cool. As they cooled, tropical forests began to retreat and were replaced

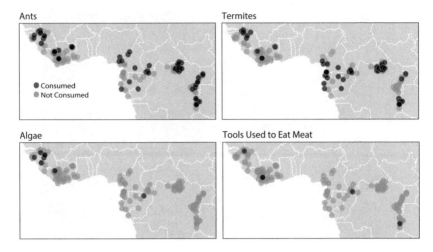

FIGURE 2.2. The map shows sites at which chimpanzees have (dark grey dots) and have not (light grey dots) been observed using tools to feed on ants, termites, algae and meat.

by woodlands. Then woodlands retreated as new kinds of grasses evolved and grasslands expanded.[30] In response, our own ancestors began to forage ever farther, from one patch of forest to the next, through grasslands (in contrast, the ancestors of modern chimpanzees appear to have stayed in the trees). Those of our ancestors who could walk more upright were better able to cross such gaps between forests.[3] They might have taken advantage of burned grasslands for easier, safer, walking (as do chimpanzees in Fongoli in Senegal).[31] All of the small number of human ancestors and their relatives that have been found from this time period show evidence of spines, hips, and feet slightly better able to walk upright, on two feet, than those of chimpanzees. These species still looked more like chimpanzees than like modern humans and yet had begun to change. It is assumed by most paleoanthropologists that during this period our ancestors began to use new kinds of utensils associated with finding foods, whether sticks for spearing (like the chimpanzees in Fongoli) or sticks for digging out roots in forest patches or along ponds and rivers.[4] Whatever these utensils were, they are either lost to the teeth of time or hiding somewhere, beneath the ground, waiting to be discovered.[5]

Around 3.5 million years ago, the forest patches our ancestors relied on became even smaller and farther apart. The climate continued to become cooler and grasslands more expansive. Grazing animals became common. Gathering enough fruit would have required traveling even greater distances. It was at about this time an animal called *Australopithecus* evolved. Species of the genus *Australopithecus* had bone structure even better suited to walking bipedally than did their ancestors. Fossils of dozens of individuals of a half dozen or more *Australopithecus* species have now been discovered.[6] They were diverse in their appearance and lifeways. Yet, they all seem to have eaten diets primarily dependent upon fruits, roots, and leaves from the forest.[7]

As much as *Australopithecus* species looked much more human than do modern chimpanzees, with bigger brains and a bipedal gait, the evolutionary changes that yielded their bodies were neither bigger nor smaller than others of the changes that had occurred in primates during the previous sixty million years. Then, around 2.8 million years ago, our ancestors and ancient relatives began to evolve more rapidly. That rapid change was associated with the origin of the species *Australopithecus habilis* (often called *Homo habilis*), but even more so with the later origin of a species of ancient human, often called *Homo erectus,* around 1.9 million years ago, as grasslands across Africa continued to expand.

Here let us pause for a second to discuss names. The names of species within the genus *Homo* are in rapid flux, as they often have been. The odds are low that the names being used in ten years, when you might be reading this, are the same as those being used now. What to do? The paleoanthropologist Chris Stringer, who thinks a great deal about such things, suggested to us a kind of linguistic cheat. All species of the genus *Homo* are humans (this isn't the cheat, this is just the reality). In terms of the story we are telling, we are often either talking about ancient humans (such as the species we called *Homo erectus* in the above) or recent humans (humans, Neanderthals, and a few other interbreeding groups of humans that lived during the last million years). To make things easier, except when we need to really call attention to taxonomy or very specific human species, we will tend to just talk about ancient humans or recent humans. Now, back to our story.

The brain of *Homo erectus*, that first species of ancient human, was about twice as big, relative to its body size, as that of a chimpanzee. In addition, it had small molars, much smaller than would be expected based on its body size, and delicate jawbones.[8] We don't know what behavioral or ecological change precipitated these changes in the bodies of ancient humans. However, a consensus is emerging that they had new ways of either processing foods or of acquiring large quantities of hard-to-find and easy-to-digest foods. It was these new food ways that allowed ancient humans to get enough energy to fuel a larger brain and to be able to afford smaller and less costly teeth and jaws. The consensus, however, does not extend to what those new ways might have been.

There exist several possibilities. Ancient humans might have figured out how to gather large amounts of honey from honeybees. Modern chimpanzees use their hands to harvest honey from honeybees, but they also use sticks. In doing so they are stung repeatedly and are limited, as a result, in terms of the amount of honey they can obtain. The same must have been true for *Australopithecus*. Eventually, however, our ancestors found ways to calm honeybees and, in doing so, to extract more honey. Smoke calms and disorients bees for ten to twenty minutes;[9] in some cases it even causes them to flee the hive. If ancient humans smoked honeybees, they could have gathered much more honey, along with brood. Plant exudates might also have been used to calm bees. Even today, tens of indigenous groups around the world use plants, either applied to their own bodies or applied to hives, to calm bees.[32] Either way, once honeybees had been calmed, honey and brood could be gathered in relative peace. They could have been gathered in large quantities. By smoking bees, Efe hunter-gatherers living today in the Ituri Forest in the Democratic Republic of the Congo are able to collect so much honey and bee brood that together these account for eighty percent of Efe calories in the wet season. It is possible that ancient humans depended in similar ways on bees; sweetness may have made big brains, small teeth, and weak jaws possible.

Or, it has been argued, ancient humans might have begun to eat shellfish. In contrast to raw mammal or bird meat, the meat of shellfish, whether they are mollusks, crustaceans, or echinoderms, is easy to digest. The collagen in connective tissue is what makes bird and mammal meat tough to chew. (It is also what helps to make mammal and bird meat juicy. The connective tissue in meat gelatinizes during cooking.) Mollusks, such as mussels and oysters, could be eaten slippery and alive the way people today eat oysters. Crustaceans, such as crabs or crayfish, can also be eaten raw. So too echinoderms, such as sea urchins.

But to have played a major role in the evolution of ancient humans, something about shellfish eating must have changed about 1.9 million years ago. Perhaps, our ancestors, at this time, innovated new ways of gathering shellfish, or of getting access to their meat. Shellfish can be hard for novice eaters to consume with ease and speed, as Brillat-Savarin noted in describing an evening in which he consumed large numbers of oysters with a dignitary. The dignitary, Brillat-Savarin wrote, "went up to thirty-two dozen," but only after "taking more than an hour over the task, for the servant was not very skillful at opening." Being skillful at opening mussels and getting the meat out of crustaceans, or at making tools to do so (paleo clam knives and crab crackers), may have been a big innovation for our ancestors. Chimpanzees do not eat mussels (at least not yet), but, as we've noted, several populations have recently been observed gathering algae; small animals are often embedded in the algae and consumed as part of the culinary whole.[33] In addition, at least one chimpanzee population has come to enjoy walking along streams, turning over rocks, and then gathering and eating the crabs found beneath them.[34] Perhaps ancient humans used tools to become more effective at such harvests.

The most often cited hypotheses about the change in lifeways of ancient humans is that they discovered how to process foods. Many of the calories in mammal and bird meat and in roots are locked up in hard-to-digest compounds. When we eat foods raw, these compounds pass through us, largely unaltered.[10] The same would have been true for ancient humans. Processing makes the energy and flavor in hard-to-digest foods more available. There are several ways ancient humans might have processed foods.

Given that chimpanzees use stones to pound and smash some of their foods, it isn't hard to imagine that ancient humans were able to do the same and, perhaps, became ever better at doing so effectively and frequently. We know that ancient humans were able to use one stone like a kind of hammer to hit another stone and produce sharp flakes as well as the leftover core (from which the flakes had come). The core could then be further modified to make another kind of tool called a hand axe (the use of which remains hotly debated). Ancient humans might have used the sharp stone flakes to cut into the food, and dull stones, including the base, to smash food. Cutting meats, particularly those of birds and mammals, makes them easier to digest. In cutting meats, ancient humans might have in essence used stone tools as stronger, sharper, more disposable stand-ins for teeth. By the time of the first ancient human (about 1.9 million years ago) our ancestors had been using stone tools for no less than 1.4 million years.[35] They might have, in that time, become pretty effective at cutting. Similarly, they might also have pounded food, as do most hunter-gatherers and some chimpanzee populations. Like cutting food, pounding food frees up energy, both by breaking off shells and husks and by macerating cells themselves such that their contents are more available. In pounding, stones, once more, serve in place of teeth.

In addition to cutting and pounding food, the first ancient humans might also have fermented foods. Fermentation is similar to cutting and pounding in that fermentation also helps to make foods easier to chew and digest. Fermentation frees up calories. It also has the added advantage that, if done right, it can kill potential pathogens. In addition, the fermentation of meat and roots can actually add nutrients not otherwise present in the food. Bacteria can produce vitamin B_{12}. They can also gather nitrogen from the air and turn it into amino acids. Unfortunately, the archaeological record is silent with regard to the possibility that ancient humans might have fermented some of their food. Recently, Katie Amato, a primatologist at Northwestern University, has begun to argue convincingly that the first human species *could have* fermented food (a possibility we consider in more depth in chapter 7), but whether they did remains entirely unknown.

Then, there is fire.

In his book *Catching Fire*, the primatologist Richard Wrangham argues that fires and cooking are *the* defining features of the evolution of the first, ancient humans. Cooked food, Wrangham hypothesizes, is what offered our ancestors enough energy to make it possible to evolve larger brains.[11] For cooking to have been the key factor that influenced the evolution of those humans, it needs to have begun no later than about 1.9 million years ago. Yet, the oldest moderately strong evidence for the controlled use of fire in cooking is much more recent. However, in fairness, evidence of 1.9 million-year-old fermentation and honey gathering are also lacking, and so too, evidence for a dramatic increase in meat and root cutting or pounding or shellfish eating.[12]

But, regardless of whether Wrangham's controversial big idea about fire is right, embedded in that idea is what we consider to be a far less controversial hypothesis. The hypothesis is not about when or whether fire shaped the evolution of our ancestors. It instead relates to why our ancestors would have innovated new food ways in the first place. It is a hypothesis that relates as much to cutting, grinding, and fermenting as it does to fire. Here and there within his book, Wrangham argues that the proximate reason that our ancestors began to use fire was because cooked food is delicious, or at least more delicious than raw food. Yes, fire can free up extra calories in food. It might even offer more free time for doing new things, things like inventing language and stone tools. But it didn't occur in anticipation of these changes. Rarely do animals, modern humans included, make choices based on long-term benefits. Instead, Wrangham argues, whenever our ancestors began to cook they did so because cooked foods taste more delicious than do raw foods. Let's think about what Wrangham's idea means for a moment. The use of fire, fire that keeps us warm and lights our way, fire that we have chained inside our ovens so as to heat food in our kitchens, fire that would make way for combustion engines, modern cities, modern war, the internet, and so much else, may have begun because it made food more delicious.

Let's give Wrangham's hypothesis a name so as to be able to remember it. Let's call it the *flavor-seeker hypothesis*. The flavor-seeker hypothesis applies to the role of fire regardless of when fire was first controlled. Wrangham does not need to be right about the importance of fire to hominin evolution for the flavor-seeker hypothesis to be right. The hypothesis simply says that whenever fire was used, it was used, first and foremost, because cooked food was more pleasurable and flavorful to eat than the alternative. Nor does the hypothesis need to apply only to fire. The flavor-seeker hypothesis describes aspects of the culinary traditions and cuisines of chimpanzees. Chimpanzees make and use tools to seek out flavor, and the specifics of the tools they use relate in part to the environments and in part to their traditions. The flavor-seeker hypothesis also has the potential to explain why our ancestors began to employ other food-processing technologies, whenever they did so. But the flavor-seeker hypothesis makes a big assumption. It assumes that the foods our ancestors used new tools and techniques to access were actually more flavorful than were those that were otherwise available. Most evidence suggests they were.

As we've noted, chimpanzees have taste receptors and preferences similar to those of humans. This is true despite the radical differences in the diets and bodies of chimpanzees and humans. Our large intestines have evolved differently than theirs (ours are shorter and less able to digest leafy greens). Our mouths have evolved differently (smaller teeth, weaker jaws). Our stomachs have evolved differently (ours appear to be much more acidic, though the data are few). Our digestive enzymes are even different than theirs, or some of ours are, anyway. Some adult humans have versions of genes that allow them to continue producing the enzyme lactase long after childhood. The persistence of lactase allows many adult humans to drink and digest milk long after weening. The same is not true of chimpanzees or any other mammals. And yet, despite our differences, our human taste receptors seem very similar to those of modern chimpanzees and, in all likelihood, our common ancestors. They are similar despite the combined twelve million years our bodies have had to evolve since our divergence from each other six million years ago.

Human sweet taste receptors and umami taste receptors appear to be very similar to those of chimpanzees. Chimpanzees also seem to be attracted to salt and even acidity (sourness) at concentrations similar to those humans find to be appealing. Studies in zoos find that chimpanzees and gorillas, once they get over novelty, tend to like foods in more or less the same rank order that zookeepers or other humans might like them. All prefer a mango over an apple and an apple over an uncooked potato.[36] All prefer foods that are rich in the chemicals that trigger the umami taste to those that don't. All seek out salt (even if they have too much already).[37] As a result, the tastes of the foods that chimpanzees eat (as perceived by humans) may be a reasonable proxy for some of the tastes that would have been available to our common ancestors living in forests six million years ago. This reality is useful for testing the hypothesis that our ancestors were flavor-seekers, and it may, in fact, be a consequence of the extent to which our ancestors were flavor-seekers.

Most researchers who study chimpanzees try at least some of the foods the chimpanzees eat. It is hard to avoid. You follow and watch the chimpanzees for hours and hours, and you wonder, "What does that taste like?" The temptation is even greater if you happen to be hungry. It is greater still if you are hungry and the chimpanzees appear to be enjoying whatever it is that they are eating. You pop a ragged bit of fruit in your mouth and have the answer and, if you are lucky, some satisfaction of your hunger. As Daniel Lieberman, the author of *The Story of the Human Body*,[38] noted in an email, "It is really fun." In 1991, the chimpanzee researcher Toshisada Nishida began a more systematic taste test. It was to be a six-year effort to follow nine male chimpanzees though the Mahale Mountains at the eastern edge of Lake Tanganyika in Tanzania. The chimpanzees climbed with relative ease from tree to tree. Nishida stumbled below them as fast as he could go. From amongst the diversity of plants used by Mahale chimpanzees, Nishida was able to taste 114 different plant-based foods eaten by chimpanzees.[13]

Some of the foods eaten by the chimpanzees were bitter to Nishida. This was the only taste that didn't tell Nishida very much about the experience of the chimpanzees. It is known that bitter taste receptors of humans and chimpanzees are somewhat different. The plants that were bitter to Nishida might not have been bitter to the chimpanzees, or to our common ancestors.[14] But what about the sweetness, sourness, savoriness and saltiness of the fruits, fruits that benefited by attracting chimpanzees and other dispersers to eat their pulp and disperse their seeds? On average, the fruits didn't taste very good. Quite a few of the chimpanzee plant foods that Nishida ate were edible but bland. He described such foods as "insipid," from the Latin *insipidus*, without (*in*) taste (*sapidus*). Other primatologists have described the dominant flavor as "mealy." Hungry people eat tasteless, mealy foods; hungry chimpanzees do too, foods that are edible and yet hard to love. This is the same conclusion that Richard Wrangham came to with regard to the fruits at Kibale National Park in southwestern Uganda. The fruits available to chimpanzees appear to be particularly bland during the dry season.[15] In other words, the chimpanzees of Mahale and elsewhere in Africa were born not into an Edenic garden of flavors, where each fruit was perfect and wondrous, but somewhere that was, on many days, duller.[16]

Tellingly, however, the chimpanzees chose the fruits that were sweet or even sweet and sour when they were available, fruits such as a fig species in Mahale which Nishida described as being like a fig that one finds in a Japanese market. It had an agreeable perfume and, as Brillat-Savarin might have put it, a sensation of "tart freshness." Subsequent research has shown that the chimpanzees often remember the locations of such favored fruits and the timing of their fruiting, and head to them when they expect the fruit to be ready. They make a beeline through the trees toward sweetness.

It seems reasonable, on the basis of these studies, to imagine the common ancestors of humans and chimpanzees living in the forests of their origin, seeking out sweet, savory, and sweet and sour fruits. They wouldn't have always been able to find them, but when they did, they were pleased. They remembered. They remembered both where they'd found the fruit and when they found it, so that they could return. Eventually, they began to use tools to find new foods. These foods ultimately offered them calories, but their immediate reward was flavor. One piece of evidence in favor of the idea that it is flavor, not need, that comes first is that some of the foods that chimpanzees use their tools or other forms of ingenuity to obtain do not actually appear to be worth the effort nutritionally. Yet, they taste good.

Being a gastronome comes with trade-offs. Brillat-Savarin wrote that "the gastronome is perhaps a kind of fool, getting excited about things of no importance." We might rephrase his assertion slightly to say that, with regard to survival, a gastronome *can be* a kind of fool. The sweet, salty, and umami taste receptors of primates exist because, on average, they tended to point them toward their needs. In seeking out their favorite tastes and flavors with tools, chimpanzees are gastronomes, which is to say they take pleasure in eating. So too the ancestors we share with chimpanzees. When gastronomes use a tool to find a food that both is flavorful and provides needed calories or nutrients, gastronomy pays off. Such types of tool use might be particularly likely to be passed one generation to the next for the simple reason that individuals engaging in this use are more likely to survive. But other types of tool use might yield better flavors without necessarily yielding better nutrition or more calories. For example, some chimpanzee populations expend great effort to gather ants. The ants are tasty (as attested by the many people around the world who eat the same kinds of ants). At Mahale, Nishida found that the chimpanzees spend one to two percent of the daylight hours fishing for ants, but because the ants offer so little to the chimpanzees nutritionally, he was left with the conclusion that "the adaptive significance of this tool-using behavior is (. . .) unclear."[39]

Another case of the gastronome as fool relates not to chimpanzees, but instead to gorillas. Trees of the species *Pentadiplandra brazzeana* produce fruits containing a protein that short-circuits the sweet taste receptors of mammals. The protein is perceived to be a hundred times sweeter than sugar, so the plant needs to use very little of the stuff to attract mammals. It is cheap to produce, so a boon for the plant. But it provides almost no calories and so is of little value to the mammals that eat it. But the mammals don't know any better, so season after season they gather the red, sweet-seeming fruits, eat them, and disperse their seeds. The one exception is gorillas. Duke University scientist Elaine Guevara and colleagues discovered that all gorillas have a mutation in their sugar taste receptor gene that prevents the *Pentadiplandra brazzeana* fruit from tasting sweet. Guevara was able to show that once this mutation evolved, it spread fast, ultimately becoming the universal version present in gorilla populations. To spread in this way, this version of the gene had to, in some way, be very advantageous. It was very advantageous because individuals with the gene didn't waste their time eating fruits that were not nutritious. The implication is that gorillas had been eating so many of such fruits that it was decreasing their well-being. They had been suffering from a gastronomic foolishness that could only be remedied by evolutionary change.[17]

In many cases it seems likely that the benefits of seeking flavors were conditional. The use of a tool to gather honey might pay off in some cases, but not in others. For example, chimpanzees in many populations use sticks to gain access to honey in honey bee and stingless-bee hives along with the bee brood in those same hives. The honey is far sweeter than any fruit in the forests in which chimpanzees live. The bee brood has a fatty mouthfeel and a salty, umami-rich taste. Honey is full of energy; bee brood is full of fat and protein. For the honey-gatherer, deliciousness is often rewarded with nutrition. Yet, sometimes chimpanzees appear to spend more energy getting honey than they get from the honey in return.[18] In other cases, honey-seeking chimpanzees might acquire energy but at the expense of other nutrients, much as often happens with modern humans. This has become more likely as the habitats in which chimpanzees live have begun to change, as they have in the Bulindi forests in Uganda. The chimpanzees of the Bulindi community are now living in a landscape that has tiny patches of forest surrounded by an immense landscape of orchards and small farms. In this setting, the chimpanzees must choose between their culinary traditions and the new choices. They choose new. They find mangos and eat as many as they can. They enjoy sweet, fatty jackfruits until round-bellied and pleased.[40] They also eat guavas, papayas, bananas, passionfruit, and even the pulp of cacao fruits.[41] Meanwhile, in the nearby communities of Kasokwa and Kasongoire, chimpanzees live in an even more altered landscape. Instead of fruit trees, they find, at the edge of their habitat, sugarcane plantations in every direction. In amongst this sea of farms the chimpanzees have discovered that farmers often pile cut sugarcane at the edges of their fields. The chimpanzees sit in those piles eating slightly rotten, slightly sweet sugarcane for many hours a day.[19] In doing so they might be finding the best nutrition they can get, given the disruption of their habitat, but they might also just be being "foolish," giving in to the pleasures of sweetness for hours and hours on end, eating the sugarcane for no other reason than that it tastes good to them. It tastes so very good to them.[20]

We argue that the common ancestors of chimpanzees and humans, like modern chimpanzees, were gastronomes. They chose foods and sought out foods based on their deliciousness. As their climates dried and the flavors available to them become duller, they would have had an incentive to make and use ever newer kinds of tools. With tools, the flavors available to them included those of crunchy ants[21] and fatty termites, as well as

FIGURE 2.3. A female chimpanzee watching primatologist (and photographer) Liran Samuni while heartily enjoying figs, *Ficus mucuso*, in Budongo Forest in Uganda.

honey and bees. But they also found other delicacies, including the even larger amounts of honey available once bees could be calmed. If modern chimpanzees and modern human hunter-gatherers are any indication, our ancestors would have loved honey, especially once it could be obtained in large quantities. For example, Colette Berbesque at the University of Roehampton recently asked a group of Hadza hunter-gatherers which foods they liked. She found that both male and female Hadza regard honey as the tastiest of foods, tastier than berries, tastier than baobab fruits, and even tastier than meat.[22] The Hadza individuals that Berbesque interviewed said they gather honey because it is delicious. It might seem obvious that modern hunter-gatherers eat sweet foods because their tongues reward them with pleasure for having done so and that our ancestors might have eaten sweet foods for the same reason. Yet, this possibility has only recently crept into the anthropological

FIGURE 2.4. Honeybees ripening their nectar. Honeybees concentrate the nectar in steps. First, they swish it back and forth in their mouths and make little bubbles of nectar from which some water evaporates. Then they spread it on their honeycombs, as you see here, and fan it to evaporate more water. In addition, the bees add enzymes from their mouths to the honey. The main sugar in nectar, sucrose, does not dissolve well in water at high concentrations. It crystalizes (and becomes less useful to the bees). Think sugar cubes. The same is not true of the smaller sugars into which sucrose (a disaccharide) can be broken, glucose and fructose (two monosaccharides). European honeybees have an enzyme in a pouch in their heads that helps them to take advantage of this biochemical reality. The enzyme, which the bees apply to the concentrated nectar, breaks most of the sucrose into glucose and fructose and, in doing so, allows honey to become even more concentrated, indeed the most highly concentrated sugary substance in nature. The sugar in honey is so concentrated that bacteria that try to feed on it die. Their cells try to balance the water inside the cell and outside the cell and shrivel, their water having left them for the sugary outside world.

literature. As a result, when ethnographers note that hunter-gatherers eat what pleases them, they typically do so with what almost sounds like surprise. Colette Berbesque, for example, writes: "It is interesting that men bring back to camp more calories of honey per hour gone from camp than any other food type, followed by meat, baobab, berries, and finally tubers. This happens to be exactly the same as their preference ranking. This suggests that men might devote more effort to getting the foods that they like the most."[42] Berbesque has to write this circumspect paragraph arguing that maybe, just maybe, hunter-gatherers eat things because they are tasty, because it is an idea of which her colleagues need to be convinced.

As for cutting, pounding, fermenting, and cooking, they too improve flavor, especially mouthfeel. Mouthfeel is a key component of flavor; mouthfeel can be silky, coarse, tender, or smooth. But it can also be chewy or stringy. Raw, unprocessed warthog, rabbit, or even elephant meat can be, it is said (typically with a grimace), "palatable," but they don't have a good mouthfeel. They are chewy and stringy and hard to swallow (especially, as the primate behavioral ecologist Hjalmar Kuehl has pointed out, for older animals with missing teeth). Uncooked meats, as Harold McGee notes in his book *On Food and Cooking*, have a "kind of slick, resistant mushiness." Chewing big pieces of raw mammal meat, McGee goes on to point out, compresses it instead of cutting through it. It is slippery and unsatisfying.[23] The food writers Hsiang Ju Lin and Tsuifeng Lin, authors of *Chinese Gastronomy*, put it even more directly: "Raw fish is insipid, raw chicken metallic, raw beef is palatable but for the rank flavor of blood."[43] Similar complaints can be lodged against the experience of eating many kinds of roots raw. Think of subsisting on raw potatoes or raw cassava. Many words come to mind to describe the experience; none of them are "delicious." A few kinds of plant roots, including those of carrots and radishes, are tasty, but they are the exceptions rather than the rule.

Cutting, pounding, fermenting, and cooking all soften foods and make them easier and more pleasurable to chew. Roots that have been smashed, cut, fermented, or cooked are easier to chew. So too is raw meat. In addition, cutting raw meat would have allowed our ancestors to divide animals into different cuts of meat, some of which would have been more pleasing to eat raw than others. Cutting, cooking, and fermenting together would have allowed soft bits of animals, such as their stomachs, to be eaten raw, hard pieces to be cooked, and still other pieces to be fermented. Once our ancestors could smash, chop, ferment, and cook, they could (and would) have availed themselves of varied techniques for different bits. Chimpanzees spend forty percent or more of their waking hours chewing their food. They chew the pulp of fruits. They chew leaves. They chew insects, meat, and, in some communities, roots. In contrast, it is estimated that once our ancestors began to cook, they needed to spend much less time chewing, perhaps as little as ten percent of their day (today the average human spends about 4.7 percent of the day chewing). This reduction in chewing time offered our ancestors more time and energy that could be used to imagine new kinds of tools, forage for riskier foods (that were very tasty, but not very predictable), care for their children, make art, or invent jokes.[44] Ancient processed food had a better mouthfeel in and of itself *and* left more time for other pleasures.[24] Mouthfeel and these other pleasures may have been the main proximate advantages of smashing and cutting foods.

Meanwhile, the idea that tools might have been used to access the flavors of shellfish is unsurprising to anyone who enjoys the flavor of raw oysters. Cracked free of their shells, the muscle of mussels and the flesh of crabs have a pleasing texture. They have more umami tastes than are present in most forest foods. In addition, they are easy to chew. Mussels, especially young mussels, don't actually even need to be chewed. What Brillat-Savarin said about French banquets and oysters might just as well have been said about ancient hominins and other mussels.

> I remember that in the old days any banquet of importance began with oysters, and that there were always a good number of the guests who did not hesitate to down one gross apiece (twelve dozen, a hundred and forty-four).

Several relatively early hominin sites have now been discovered at which extraordinary quantities of mussels appear to have been eaten, as evidenced by the shells that remain. It is, as of yet, unclear whether these piles represent the eating of a few mussels a year for many years or many mussels at a time. But at least some of our ancestors may have had, to use Brillat-Savarin's phrase, "a bellyful of them."

Meanwhile, cooking and fermentation do more than just alter the mouthfeel of foods. They also alter and improve their tastes and aromas. As meat or roots are fermented or cooked, the amount of free glutamate present in them increases dramatically engendering umami tastes. In addition, as we'll consider in the next chapter, the aromas of the meat become much more complex; this complexity may be instinctively pleasing. Similarly, as roots are cooked their complex carbohydrates begin to break down and their simple sugars begin to caramelize. A raw sweet potato is scarcely food. A fire-roasted sweet potato, crisp on the outside, sweet and soft inside and redolent with pleasing aromas, is something worth telling a friend about.

Modern apes, like modern humans, appear to prefer cooked food. In zoo environments, chimpanzees and gorillas both choose cooked vegetables over raw vegetables when offered both. Chimpanzees also choose cooked meat over raw meat (gorillas don't eat meat).[45] Every indication is that chimpanzees and gorillas make these choices because of the flavors of cooked foods.[25] The apes have as much as said so. In a telling experiment described by Wrangham, Koko, the gorilla who had been taught sign language, was asked by psychologist Penny Patterson if she preferred cooked vegetables, which Patterson held in her left hand, or raw vegetables, which Patterson held in her right. Koko touched Patterson's left hand, the hand containing the cooked vegetables. Koko was then asked if the reason she preferred them was because they were "easier to eat" or because they "taste better." Koko indicated "taste better," as likely would have our ancestors. No one has tested whether the apes that eat some meat—chimpanzees and bonobos—also prefer fermented food to raw food (except in the case of alcohol, where the effects of flavor and those of drunkenness are hard to disentangle).

Backing up a little, we suspect that the first ancient humans used their big brains to engage in a number of new food-finding and food-processing behaviors. They were guided by the search for delicious foods, which most of the time were also nutritious foods.

As our ancestors wandered, they used their ever-bigger brains to seek out tastes and other flavors and, in doing so, find much needed nutrients. The better they became at this seeking, the less relevant the various parts of their digestive systems that had evolved to process food became. Big teeth and jaws help tear and grind food. The teeth shrank. Jaw muscles became smaller. The large intestines help break down complex compounds into simpler-to-use forms. They too, in other words, process food. The large intestines became shorter. Each of these changes was almost a kind of atrophy, atrophy made possible because these parts of the body were no longer as necessary to survival. Cave fish, over enough time, lose their eyes. Our ancestors, after enough time eating food that was energy rich or processed, food made available thanks to the search for flavors, lost elements of their guts, teeth, and jaws. Natural selection no longer acted quite so strongly on these parts of the body; and to the extent to which it did act, it acted to winnow them and reduce the energy wasted on their production and maintenance and use that energy instead on the ever-growing brain.

Around 1.5 million years ago, ancient humans began a multi-generational walk across Africa and on into Asia and Europe. We can't know why they spread. We hypothesize that the list of reasons included search for food (and hence flavors). They spread from regions where food was less abundant or delicious to regions in which it might be more so. They spread from hill to valley around half of the terrestrial world. As they did, they met new challenges with what University of Arkansas paleoanthropologist Peter Ungar has called "dietary versatility," a kind of mouth-first adventurousness.[46] Or to paraphrase Wordsworth, "homeless near a thousand homes" they stood and near a thousand fires they "pined and wanted food." They traveled and pined for and wanted food and, as they did, diverged into half a dozen or more different species or lineages, some of them isolated in particular regions. Each of these lineages carried with it the knowledge necessary to make a full set of utensils wherever they went, much as a modern chef might carry her knives, a full set of utensils, and at least some knowledge of how to process food so as to make it more flavorful. Recently, it has become popular to try to eat as our paleolithic ancestors might have eaten. But such attempts inevitably run into a rather central problem: ancient humans ate different things each place they went, in some places shellfish, for instance, in others the marrow and grease of bones.[47]

As the Stone Age progressed, these differences from one population to the next would become even greater. Like modern chimpanzees, the diverse recent human species each had culinary traditions and cuisines. But given their enormous geographic range, these traditions would have been even more pronounced, if for no other reason than that the habitats ancient humans inhabited ranged from rain forests to tundra, from the Congo Basin to what is now continental Europe. These humans appear to have adapted to their many different surroundings primarily through finding new ways to survive and eat in each place they found themselves. Of course, genetic evolution occurred too. It was the combination of cultural and genetic evolution that would eventually lead to the origin of the cavalcade of recent human lineages,[26] including Neanderthals (*Homo neanderthalensis*), Denisovans, and us (*Homo sapiens*). Yet, this is what is remarkable: the taste receptors of these species remained very similar. We know this from direct comparisons of the genes. Recent studies based on ancient DNA extracted from teeth and bones have found that the sweet and umami taste receptors of Neanderthals, Denisovans, and our own species, *Homo sapiens*, are nearly identical to each other. Meanwhile, to the extent that bitter taste receptors differ, they simply broke in slightly different ways in different species; they broke, it has been suggested, because our ancestors had found ways to make some of what was dangerous safe.[48][27]

As these recent human species, divergent in their body forms and geographies, tried new foods and new ways of processing foods, they learned to distinguish the pleasurable from the bland and the safe from the dangerous. They did so again and again. Their tongues would have been a help during this process, a starting point for understanding. Their tongues told them what was bitter. Their tongues told them what was sweet. But the tongue did not act alone. The nose was also a guide. Once humans became able to process foods, they needed a way to remember which forms of processing were safe and which were not (a lesson typically learned the hard way). They also needed to remember which foods that seemed dangerous based on their taste were actually safe. In this context, the nose came to take on a more significant role. There were no libraries, nor any way of storing information beyond one generation other than what might be accommodated in oral histories. This would be true for an extraordinarily long period of time, from the origin of humans until eight thousand years ago. Across these long years our ancestors depended disproportionately on their noses. In this they were not alone. A mouse uses its nose. A dog uses its nose. A pig uses its nose. But human noses are different; humans would come to rely upon their noses in entirely novel ways. They came to rely upon their noses to catalogue flavors, ranking them from delicious to deadly and responding accordingly. This is a reality made most clear in consideration of an unusual, wonderful, modern food, the truffle.[28]

CHAPTER 3

A Nose for Flavor

First parents of the human race [. . .] what would you not have given for a truffled turkey hen? But in your Earthly Paradise you had no cooks, no fine confectioners! I weep for you!

—JEAN ANTHELME BRILLAT-SAVARIN,
THE PHYSIOLOGY OF TASTE

Bees are led by odor of honey, vultures by carcasses. . . . The powerful scent of hunting dogs, sent on ahead, will lead the hunter in whichever direction the cloven-hooved wild beast has gone.

—LUCRETIUS, *DE RERUM NATURA*

We've intentionally left out one piece of the story of flavors until now: aromas. We've left aromas out not because they are unimportant but because they are so important that they require separate consideration. As Brillat-Savarin noted, "There is no full act of tasting without the participation of the sense of smell." This is especially true for humans. It is in light of the uniqueness of the human sense of smell that some of the major transitions in human evolution, including the control of fire, become most clear.

Smell is elaborate and consequential and, more so than taste, it differs among species. One element of the difference relates to acuity. Species differ in terms of the concentrations of compounds that they detect. But species also differ in terms of the relationship between the nose and the mouth, between scent and flavor. It is helpful to understand these distinctions in order to understand the unique role that smell plays for humans. These distinctions can be readily seen in light of a modern, foraged, food: truffles. Our ancestors once had to search for all of their food. Truffles are a reminder of that time and way of life. Where they grow, truffles are the emblem of that which is wild and delicious and yet must be found. In France, Saint Anthony, the patron saint of lost things, is also the patron saint of truffles. He will help you find your keys or your truffles, but with regard to the latter it also helps to have a dog, or even better, a pig.

Each species that searches for truffles perceives them in a different way. For pigs, the attraction to the aroma of truffles is innate. For dogs, the attraction to the aroma is learned. For humans, the attraction is partially learned and maybe also partially innate and relates primarily to the aroma of the truffle once it is food and inside the mouth.

The roots of the story of humans, pigs, dogs, and truffles are old. For no fewer than a thousand years, and potentially far longer, the French have used pigs to hunt truffles in the forest. A pig can find a truffle even if it is growing a foot beneath the ground. When we use an animal to find something we can't find on our own, we can take advantage of the differences between that animal's senses and our own. In the case of the pig there is an added component: pigs are innately attracted to truffles, drawn to them by a kind of biochemical gravity.

FIGURE 3.1. A pig's snout.

Truffles are produced by very specific kinds of fungi. These fungi, including species of the genus *Tuber*,[1] form relationships with the roots of specific tree species. Truffles of the genus *Tuber* partner with roots of beech, birch, hazel, hornbeam, oak, and other trees. The fungi exchange signals with the tree roots using the primordial language of biochemistry and connect to them at the molecular level. The relationship is ancient and intimate and has evolved repeatedly over millions of years.[49] The roots of trees are relatively thick, and so it is difficult for them to find nutrients deep in the interstices of the old, rocky soils. The root-like hyphae of fungi, however, are far thinner and so can tap into pockets of water and nutrients the tree roots are unable to reach. The density of these hyphae can be great. A single cubic inch of soil might contain a mile of fungal hyphae. In exchange for offering the trees use of what they gather through this vast web of filaments, the fungi levy a tax of sugar.

The relationship between tree and truffle is a mutualism of exchange akin to those that the very first land plants are thought to have forged with fungi in order to colonize, mile by mile, the entire ice-free Earth.[50] Many fungi form partnerships with the roots of trees, but the fungi that make truffles are especially good partners. In addition to providing their trees with nutrients, they also release compounds into the soil that kill other plants, thereby eliminating competition their favored tree might have. This creates a brown zone around trees with truffle fungi, a zone the French call the *brûlé*, the burn.[51] But what makes truffle fungi really very special is that they have evolved a unique way of sending their progeny off to colonize new tree seedlings (seedlings that in some cases do not grow well, or even at all, until colonized). They do so by making truffles.

Truffles are an underground mushroom, a fungal fruit. They are produced when the fungi of one sex meet up with those of another sex and mate. Truffles look like a knobby, earthen brain. They are hard, not soft, and can be large or small, but generally speaking one counts oneself as extraordinarily fortunate to find a truffle the size of a walnut. The spores of the truffle reside in masses of tissue in and among the "brain's" folds. Like any mushroom, the truffle succeeds if it finds a way to get its spores as far away as possible, to some new ground where they too might thrive, ideally far enough away so as to avoid competition with their parents. Many mushrooms do this by attracting animals that consume and carry their spores. The truffle's challenge is that it is underground. To overcome this challenge, truffles must be, as Merlin Sheldrake puts it in his book *Entangled Life*, "pungent enough for their scent to penetrate the layers of soil and enter the air, distinctive enough for an animal to take note amid the ambient smellscape, and delicious enough for that animal to seek it out, dig it up, and eat it."

There are many species of truffle-making fungi. Each achieves its mixture of pungency, distinctiveness, and deliciousness in a different way, different because each truffle species appears to have evolved to woo different animals. The truffles that humans most love evolved to woo pigs, and they do it well. The response of pigs to the aroma of truffles appears to be, at least partially, genetically encoded. Even a very young pig, naïve to the world's stinks, is drawn to the aromas of truffles. Truffle aromas are inhaled up the pig's nose. Once there, they contact receptors inside the nose. The receptors wave around like sea anemones. Much as sea anemones sift the sea for nutrients, these receptors sift the air that comes into the nose for airborne chemicals. Each receptor is tuned to one or more particular airborne compounds. The compounds are the keys in the locks of the nose's receptors. Once "inserted," the key leads the receptors to trigger nerve cells in the olfactory bulb, which then relays messages to the brain.

A very small number of aromas appear to be hardwired in the brains of mammals in ways that lead to fear (and disgust) or pleasure. For example, a number of predator odors trigger fear in mice. Long before a predator arrives in a patch of forest, the aromas that precede it reach the olfactory receptors of mice, young and old, and signal their brains, "Run. Hide. Freeze. TERROR. RUN!" The chemical 2,5-dihydro-2,4,5-trimethylth iazoline is reliably present in the feces of foxes and wolves and triggers fear even in newborn mice whose families have lived in the laboratory for generations and never encountered a canid.[52] Another compound, present in cat saliva and spread to cat fur when cats groom, has a similar effect. Similarly, many animal species appear to instinctively avoid the chemicals putrescine and cadaverine, associated with the decay of vertebrate bodies.[2] This response appears to include a conscious emotional component, that of disgust. It also includes unconscious behavior changes. Humans who have been experimentally exposed to putrescine, for example, are more alert and aware. This is true even if those humans were not consciously aware of being exposed. It is hypothesized that being alert around the smell of putrescine might help animals to anticipate dangers. Its aroma tells our brains, "Something bad happened here. Look up. Turn back."

The aromas that are hardwired to be displeasing or fear-evoking are typically associated with danger. Those that are hardwired to be pleasing are most often associated with sex. For obvious reasons, it is advantageous for animals to be hardwired to be pleased by the sex pheromones of their own species. Female Asian elephants, for instance, release the compound (Z)-7-dodeceynl acetate in their urine to signal to males that they are ready to mate. The male elephants respond instinctively to this sexy urine.[3] Female goats begin to ovulate upon smelling the aroma of male goats, an aroma that is wicked out through the fur on their heads (and often ends up in their meat).[53] Similarly, male boars produce androstenol and androstenone in their testes. Androstenol smells musty (at least to humans). Androstenone smells "urinous." From the testes of a boar, these compounds migrate all the way through the boar's body to a special salivary gland. When a boar is in the mood, the contents of this gland leak into its mouth. The boar then aggressively chomps its jaws, shakes its head and snorts. If things are going well, it does so in the direction of a sow. The sow responds instinctively to the aroma of the boar's libidinous and frothy spit and assumes a mating stance.[54] Instinctively pleasing aromas aren't, however, just sexual. For example, cadaverine, the namesake fragrance in death's boquet, attracts and pleases vultures, carrion beetles, and a whole zoo of death-loving flies. What disgusts one species can entice another. Indeed, what disgusts one individual within a species can disgust another within the same species.

Truffles produce a chemical perfume to attract the mammals they need. The truffle aroma includes androstenol, one of the two steroid-like compounds required to cause female pigs to assume the mating stance. The truffle aroma also includes dimethyl sulfide; dimethyl sulfide smells like slightly rotten cabbage.[4] Dimethyl sulfide is enough to attract a pig to a truffle and has the advantage for the truffle of being detectable at very low concentrations.[55] We might imagine a pig walking toward the aroma of stinking cabbage (dimethyl sulfide), only, as it gets closer, to detect ever more hints of sexy pig (androstenol) until it is right above the truffle and begins to dig. We cannot know for sure whether, upon finding a truffle, a pig thinks about sex, food, or some complex feeling in between. We know only that the pig is pleased.

Today, dogs are more likely to be used to hunt truffles than are pigs. Dogs must be taught to find truffles; they do not naturally desire them. This is the advantage of using dogs. It is easier to reward a dog for finding a truffle it has learned to seek than it is to wrestle a truffle from a pig that thinks that it has found something wonderfully edible and sexual all at once.

Recently, we went with our kids on a truffle hunt in the Dordogne region. The Dordogne is in the southwest of France, just east of the city of Bordeaux and north of Toulouse. A dog would do the actual hunting, and we were to follow along. The best truffles in the world are said to grow in the Dordogne (unless you are Italian, in which case they, obviously, grow in northern Italy). We did not travel there for the truffles. Instead, we had come because it is in this region the caves visited by Neanderthals and *Homo sapiens* are densely clustered. The first Neanderthals were present in the Dordogne by no later than two hundred thousand years ago. Our own species, *Homo sapiens*, arrived more recently, having evolved from one or another descendant of ancient humans and spread via the Middle East, some forty thousand years ago.[56] As for our family, we showed up in 2018.

Neanderthal populations were never very dense in the Dordogne. Yet, the Neanderthals lived in the region so very long, more than four hundred thousand years, that the soil is full of their bones and tools. Meanwhile, the bones and stones of the earliest *Homo sapiens* in the region are even more ubiquitous. By 30,000 years ago the populations of *Homo sapiens* in France might have been tenfold those of Neanderthals at their peak.[57] Eventually, some of those *Homo sapiens* began to make art. The results are extraordinary. The caves painted by the paleo-humans of the Dordogne are something akin to a prehistoric Louvre, filled with hand prints, finger tracings, ancient symbols, magical humans, and dynamic scenes of mammoths, woolly rhinoceroses, and horses. The two of us are drawn to this art, like pigs to truffles. To be deep in a cave and see art carved, painted, and blown on walls by artists tens of thousands of years ago is to be moved in a way that is only partially conscious. It was this art that we had actually come to the Dordogne to see. As they say, we came for the stones and cave art, but we stayed for the food and wine. Maybe they don't say that, but that is what we did, and where we happened to stay was in a small hamlet in which Edouard and Carole Aynaud, after lives of traveling the world, have decided to dedicate their retirement to growing and hunting truffles.

We went out to the orchard behind their house on a particularly lovely Sunday morning. We were joined by a dozen or so other truffle hunters. Together, we followed Edouard and his dog. The dog had been trained to hunt for truffles. Edouard would lead the dog to various areas of the orchard and let it sniff. Edouard made clear that we might find no truffles. It was early in the season. It was possible that no truffles were yet mature enough to be sniffed out by the dog. But it didn't really matter. We could have found no truffles and would have still been delighted. Rob has spent much of his career searching for rare species in forests. Most of the time the species, be it a beetle that rides on the backs of ants, cowboy style, or a rare bee that ferments nectar into a kind of beer, is not found. It is much better to find your quarry after having failed to find it a few times. Or that is at least what we prepared to tell ourselves. It is the same sort of thing that paleolithic human hunters that came home empty-handed might have told their spouses too (NO, I didn't kill anything, but let me make you some cave art).

The place in which we walked with the dog was just a few miles from a small cave we had explored on our own the day before. It was land over which diverse humans have been roaming, searching for their favorite foods, for hundreds of thousands of years. For the last twenty thousand years or so they would have searched with dogs by their sides.[5] Twenty-six-thousand-year-old footprints have been found four hundred kilometers to the east in Chauvet cave. The footprints are of an eight-to-ten-year-old boy walking alongside either a dog or a young wolf. These footprints have not yet been very well studied, but they suggest an ancient relationship, a relationship in which two species together, wolves and humans, were ultimately able to see, smell, taste, and accomplish more together than either could on its own. Like the boy, our footprints tracked those of the dog, in our case the truffle dog. As we followed behind, we imagined that when we came near truffles, we might smell something, a hint of what the dog smelled. Yet, in and among the trees, we smelled nothing that might suggest a truffle. We breathed in deeply, struggling to catch a whiff. There was the odor of leaves rotting, the smell of green leaves on branches, and a hint of the cows down the hill in the valley. But no truffle. Then the dog paused, in the very spot in which we smelled nothing, and began to dig. Just like that, a truffle appeared beneath its feet. Edouard Aynaud moved the dog out of the way and allowed our son to push a trowel into the ground and pull the truffle out. There it was, perfect, dark, and bulbous. We bent down and then, and only then, we were able to smell the aroma.

We don't have a photo of that moment, but what it would have shown was a dog eating a meaty treat, its reward for hard work. And, around the dog, a dozen people bent over, noses low and butts high, sniffing desperately at the truffle in our son's hand. A still life, ripe with the complexities of attraction. Later that day we would eat a truffle. It might have been the one we gathered, it might not have been (such being the subtle art of running a business to hunt for but also sell truffles). We ate it grated over noodles, with olive oil. We ate it with deep pleasure that was neither that of the pig nor that of the dog, but something else entirely.

Whatever the wiring in the pig's brain that innately associates truffle aroma with pleasure, it appears to be absent in dogs. Left to their own devices, dogs do not search out and eat truffles. Nor do wolves or, it appears, any other canids. Dogs are able to smell truffles beneath the ground, but couldn't care less that they are there. Dogs are taught to search for truffles. They learn to associate truffles with rewards, a biscuit for their troubles. Our experience of the truffle, meanwhile, is a third thing. The pig and dog, for all their differences, both sniff the truffle through their nostrils while it is still outside of their mouth. This experience of the truffle's aroma is described as orthonasal, in front of (ortho) the nose. Our human experience of truffles includes their orthonasal aromas, but is much more heavily influenced by their retronasal aromas, the aromas that rise up from our mouths into the backs (retro) of our noses.

To explain the different between how dogs and pigs experience truffles and humans experience truffles, we need to take a moment to explain the evolution of smell. The very first noses in vertebrates evolved in a fish something like the modern lamprey (a kind of sucker-faced fish that today lives in both coastal and freshwater). The first lamprey-like nose was a pocket without an exit. Inside the pocket was a layer of olfactory (aroma) receptors, each extended on a kind of waving stalk. The early lamprey would have had just a handful of kinds of aroma receptors and so been able to distinguish correspondingly few aromas (all drifting through the sea rather than through the air). Yet this simple system was the prototype on which all other vertebrate noses would be elaborated. With time, a descendant of that lamprey, something more like a modern fish, evolved a proper nostril (the ancestor of your nostril). That nostril allowed water to enter the nose, but there was also a second hole so that the water could also exit. This flow-through system made it easier for the fish to smell the sea as it swam, each new moment bringing new aromas (in contrast to the lamprey's pocket nose, where things may well have tended to get a bit stale). As the nose became more complex, the genes that produced different kinds of aroma receptors became more diverse.[58] With each new gene came a new receptor and the ability to detect more kinds of compounds. By the time the first vertebrates crawled onto land, dragging their pot bellies and tails, they would have been able to detect hundreds of different individual aromas. By the time the first mammal, a small, long-snouted shrew-like creature, evolved, it could detect thousands of individual aroma compounds and many more mixtures thereof.[59]

Across all of these species some things have held constant. The receptors themselves, although they have become more diverse, are all of the same basic type. In addition, in each case the receptors sit in a mucous membrane inside the nose, from which they protrude. Also, in each case the receptors are tightly linked to nerve cells that lead directly to the base of the brain, to the olfactory bulb, an ancient, primitive part of the brain. Arguably, the first brain was nothing more than a nose, and this little olfactory bulb attached to it. From the bulb, some nerves connect to a part of the brain that controls innate behavior. Others travel farther, to more distant stops, where they generate conscious perceptions, what we think of when we think of the aroma of lavender, mint, or skunk, for instance.

All of this is just as true for a dog or a pig as it is for a human or a hedgehog, but there are also very important differences, differences that are easiest to see by comparing dogs and humans. The nose of the dog evolved so as to be specialized to detect aromas when the animal sniffs. A sniff is different from an ordinary inhalation associated with breathing; it is both deeper and more intentional. As Gordon Shepherd describes in his beautiful book *Neurogastronomy*,[60] a dog's sniff begins with exhalation. When a dog exhales, it blows air out of its nose through the slit-like edges of its nostrils; the air shoots out to the sides of the nose with high pressure, which helps to stir up the soil and dust and make settled aromas become airborne. The dog then inhales rapidly and sniffs these newly volant compounds, compounds floating in air composed primarily of oxygen, nitrogen, and carbon dioxide. It sniffs. The rate at which a dog respires increases to up to eight times per second when it really wants a good whiff. When being sniffed, the air does not go in the same way it came out. Instead, it goes into the nose through the central chamber of the nostril, such that the aromas being exhaled and those being inhaled do not mix. The dog draws in air from a sphere of about ten centimeters around each nostril, a sphere referred to as the nose's "reach." Gathered from this reach, the inhaled aroma chemicals travel up the nose and sit in the very long part of the nose in which receptors can be found. The dog nose has millions of individual aroma receptors of nearly ten thousand kinds. The dog's nose is optimized for sniffing, for knowing the world at the level of the dirt. The dog's nose points toward and inhales a world of rot and sweetness, a world of musk, markings, and anal sacks.

A dog's nose is specialized for orthonasal smell. Retronasal smell is a different story. Retronasal smell occurs primarily during exhalation, when breath is blown from the lungs over whatever is in the closed mouth, before it exits through the nose. Retronasal aromas probably play relatively little role in the dog's experience of the world. Relatively few of the volatile chemicals in the food a dog is chewing make their way into the dog's nose via its mouth. As a result, the flavors a dog experiences are dominated by tastes rather than the subtle mélange of taste along with aroma.[61] Its subtle aromatic experiences are reserved for the external world of paths and scent trails. In this, the dog is perfectly suited to finding truffles, but in no way predisposed to enjoying their flavor once they have been found.

The human nose is nothing like a dog nose and hasn't been for a long time.

Roughly 75 million years ago, the primate family tree divided in two. One branch, the Strepsirrhini, became the lemurs, bush babies, and their relatives. The other branch, the Haplorhini, would beget modern monkeys, apes, and humans. Once separated, many differences accumulated on these two branches. Some of those differences related to vision. The eyes of haplorhines evolved a heightened acuity and, in some lineages, a heightened ability to see colors (trichromatism). Along with these changes, the parts of the brain associated with interpreting signals from the eyes expanded. In concert, reliance on vision also increased in haplorhines. Many genes for particular aroma receptors went unused and, over generations, broke (they became "pseudogenized").[62] In concert with these changes, the nose shrank (*haplorhine* actually means "simple nose"). Monkey noses are smaller than would be expected for a mammal, given their body sizes. Human noses are even smaller, about ninety percent smaller by volume than would be expected given the average human body size.[63] These changes in the eyes and noses of the haplorhines in general and humans in particular led to necessary changes in the shape of the skull. [64] Harvard paleoanthropologist Daniel Lieberman argues that several of these shifts were associated with a decreased ability to smell orthonasally and an increased ability to smell retronasally.[65]

As haplorhine noses and eyes evolved, some skull bones were lost.[66] These lost bones were collateral damage associated with an imperfect process, the nuts and bolts left over after rebuilding things. One of those bones was the transverse lamina, a long bone that helps to separate the mouth from the nose, a kind of shelf between the mouth floor and the nose floor of the head. As Lieberman notes, this loss had potentially great consequences for olfaction.[67] While food in the mouths of the ancestors of monkeys and apes was chewed or even just manipulated with the tongue, it was suddenly, and to a much greater extent than in other mammal species, also being smelled. It was being retronasally mouth-smelled. Volatiles that rose up from food in the mouth immediately rose into the nose.

The loss of the transverse lamina (also called the lamina transversalis) may have led many primate species, including apes such as gorillas and chimpanzees, to evaluate their food in a new way. Dogs sniff their food, then bite it. Once they have bitten it, the dominant experience of the food is simple and driven by the small palette of sensations the tongue offers. Bitter. Sweet. Umami. Sour. Salty. Not so for monkeys, apes, and other haplorhine primates. Each bite of their food has a taste and, within their mouth, also an aroma. Together these sensations, along with mouthfeel and a few other bells and whistles, are flavor. Before this transition, flavor existed (indeed, each species had its own experience of flavor) but not in quite the way we, as humans, think of it today.

Once species of the human ancestor *Australopithecus* began to walk bipedally four million years ago or so, another set of changes occurred. Bipedal *Australopithecus* species were no longer sniffing the ground at their feet. To the extent that they were sniffing anything outside of what was in their mouths, they were sniffing the air. One sees this even with chimpanzees and gorillas, which are not fully bipedal and yet spend less time on all fours than do many primate species. Or at least Susann Jänig saw this during her PhD research in which she spent many hours at the Leipzig Zoo watching chimpanzees and gorillas sniff things. Both chimpanzees and gorillas can and do bend over to sniff things on the ground, but for items that can be picked up (and are not attached to other chimpanzees or gorillas) it is actually easier for them to simply lift the items, be they food, leaves, or sticks, to their noses. Or they touch those items (or each other) and sniff their fingers.[68] For the subset of things that seem to have the right smell, the next step is often that those items are licked. They are licked before they are ingested, and as they are licked their flavor is sampled, a flavor that includes taste and retronasal smell.

The evolution of bipedalism was also associated with changes in the orientation of the nasal passages relative to the torso (and hence air leaving the lungs). In humans, air being exhaled out of the lungs and along the length of the neck must make a right angle turn to exit via the nose. This sharp turn relates to the orientation of the nose relative to the head and to the way in which the neck is held relative to the body. In other primates, including chimpanzees and gorillas, the turn is less sharp. Daniel Lieberman speculates that the sharp turn around which exhaled air must travel during human exhalation may lead the air to bounce into the mouth and up through the nose, turbulently. Experiments on cadavers seem to bear out Lieberman's intuition.[69] In bouncing turbulently, the exhaled air might spread even more aromas from the mouth to the nose.

Finally, bipedal species also need to hold their food farther forward in their mouths while chewing and manipulating it (in front of the epiglottis) or risk choking. Lieberman argues that this allows more time for retronasal aromas to be appreciated. Such aromas could be savored while the tongue was manipulating the food, pushing it around and releasing volatile chemicals from the food's bottom and then top and then bottom again.

As both Gordon Shepherd and Daniel Lieberman have highlighted, these evolutionary changes in noses, heads, and bodies make hominid olfaction in general, and human olfaction in particular, unusual. One result is that humans are less able to distinguish aromas in the soil than a dog or a pig can. But they are far better at being able to experience retronasal aromas as part of the experience of flavor.[70] No one has written a better description of the human experience of the way that flavor combines retronasal smell, taste, mouthfeel, and other experiences than what Brillat-Savarin penned in 1825. He was writing about the modern human experience. Yet, inasmuch as our noses and mouths have changed relatively little over the last four million years (at least compared to the extent to which they had changed previously), it is also probably a reasonable description of the eating experience of *Australopithecus*, ancient humans, and Neanderthals.

As soon as an edible body has been put into the mouth, it is seized upon, gases, moisture, and all, without possibility of retreat. The mouth . . . is a cave in which aromas are trapped and seized upon. . . . Lips stop whatever might try to escape; the teeth bite and break it; saliva drenches it: the tongue mashes and churns it; a breath-like sucking pushes it toward the gullet; the tongue lifts up to make it slide and slip; the sense of smell however, oh that sense of smell, appreciates it . . . without . . . a single atom or drop or particle having been missed by the powers of appreciation.

Side by side in the forest or dining room with a dog or a pig, you and a dog or pig perceive different worlds. We miss part of what the dog and pig perceive, but they miss part of what you perceive. We flounder when asked to find truffles, but excel at savoring their flavor. Dogs are good at finding truffles, but fail to enjoy their flavor. Meanwhile, the pig innately and lustily runs toward the truffle and probably doesn't ever really know why. In this way, the truffle is a suitable emblem of some of the ways in which our sense of olfaction is unique, but also the extent to which the flavor world of each species is unique. One might argue, then, in returning to consider the story of ancient humans, that such humans were not only unique in their culinary traditions and cuisines, but also uniquely able to appreciate the flavors of foods, including their retronasal aromas.

A key question in considering aroma and human evolution is whether there are any aromas to which humans are innately drawn, or toward which we at least have some predisposition. As truffles are to pigs, what are to humans? No one knows.

It is possible that some aroma associated with roasted meat was as instinctively attractive, once in the mouths of the first humans, as the aroma of truffles is to pigs. Or, our brains might be primed to learn to love such foods, but not hardwired to love them. Harold McGee, author of *On Food and Cooking*, has noted that one feature that many of the foods enjoyed by chimpanzees, gorillas, and humans share is that they tend to have very complex aromas.[71] Recent studies of individual chemical compounds similarly note that humans, independent of their culture, ethnicity, or geographic origin, tend to find compounds that are complex to be more pleasing.[72] Perhaps our brains are predisposed to learn to like a subset of complex aroma compounds and then also complex mixes thereof. In Chinese there is a word, *nung*, which means, in the context of foods, "rich." It describes the way in which the human palate likes "convoluted tastes, one leading to another in convoluted paths," as Hsiang Ju Lin and her mother, Tsuifeng Lin, put it in *Chinese Gastronomy*. Once our ancestors discovered how to control fire, they found a way to alter foods, intentionally, so as to favor such complexity, which their brains may have been predisposed to love. And, compared to dogs or pigs, they especially loved such complexity once it could be savored in their mouths.[73]

Some aromas, such as those of the truffle, are complex by nature. Others are complex by culture, due to the ways that humans process food. Cooking meat is one of the ways that humans take something aromatically simple from nature and make it complex. At moderate temperatures, the chemicals in cooked meat that yield aromas come from chemical reactions in which the proteins, fats, and acids released from muscle cells combine, break apart, and become airborne. Meat begins to smell fruity, flowery, grassy, and nutty. At high temperatures, however, something else entirely happens, and it happens not only to meat but also to vegetables. At high temperatures, the deliciousness of cooked foods is transformed by food chemistry, in a process so magical it gets a French first name, the *Maillard reaction*.

The Maillard reaction is named for the physician and chemist Louis Camille Maillard, who reported the discovery in 1912.[6] Maillard did not study food. He was instead trying to figure out how organisms assemble amino acids to make proteins. Toward this end, he mixed and then warmed amino acids and sugars together. When he did, he found that totally new compounds were produced, compounds that smelled good. Maillard had, unwittingly, mimicked part of the process of cooking. In cooking, just as in Maillard's experiments, the mixture of amino acids and sugars under warm conditions produces new compounds. The compounds include pigments that make the surface of foods change texture and color. These are the pigments we see when meats brown, breads crust, or malted barley is baked before brewing. But the process also yields hundreds of other compounds, many of them small enough to be airborne and hence sensed by our noses.[7] The Maillard reaction is chemistry inasmuch as it is subject to chemistry's laws, but magic inasmuch as it remains both slightly unpredictable and incompletely understood.[8]

Every few years new chemical products of this reaction are revealed. It seems likely that they will continue to be revealed for many years to come; fire and fermentation are magicians that hide their best tricks. This aromatic complexity is characteristic of cooked meat, but also of those things in nature that have evolved to attract animals to eat them, such as fruits. More than six hundred aromas have been identified from cooked beef. But this complexity is rivaled by that of fruit and fruitlike fungi such as truffles. Ripe strawberries produce 360 compounds; raspberries produce 200, blueberries produce 106.[74] Perhaps, as McGee argues, we are innately drawn to complex aromas. And perhaps, as McGee puts it, "Cooking with fire was valued because it transformed blandness into fruitlike richness."[9] Cooking made meat, and also vegetables, complex. It turned the parts of plants and animals that did not evolve to be eaten into mixtures that cannot be improved upon, mixtures akin to those produced by fruits or truffles and yet nonetheless distinct.

What we do know for sure is that whatever our instinctive tendency to enjoy fruits, truffles, and cooked foods, such preferences are refined through learning. The ability of the nose and brain together to learn to love flavor is powerful. What is more, the large size of the brains of humans predisposes them to be able to catalogue the world of aromas. Gordon Shepherd has gone so far as to argue that our brains evolved their large size over the last several million years so as to better catalogue the species around us based on their aromas, particularly those associated with flavor.[10]

As a thought experiment toward this end, take a little piece of a mint leaf, crush it between your fingers, and then hold it up to your nose. Or better yet, put it in your mouth. When the lightweight chemical compounds in the leaf, such as menthol, rise into your nostrils, they contact and excite the dangling set of receptors to which they match. They punch the nose's biochemical keypad. A signal is then sent, an electrochemical message, that lights up a map in your brain. The map is situated on the surface of your olfactory bulb. These maps are literal; they can be seen. They look like arrangements of starbursts, cognitive constellations in which illuminated components of the olfactory bulb shine against a backdrop of darkness.[11]

Linda Buck, who discovered the workings of the olfactory receptors in the nose (and won, along with Richard Axel, the Nobel Prize for having done so), thinks that humans are probably capable of recognizing ten thousand different such constellations. These constellations are innate, subject only to genetic differences among individuals. The constellations of identical twins are, as near as we know, identical. However, our conscious experiences of these constellations and our ability to distinguish between them are both learned. We must learn to associate the experience, for example, of mint with menthol's constellation.

We are early in our understanding of how all of this works. As a result, it is easiest to describe the mechanisms using metaphors (metaphors that betray the ambiguities of our understanding). On the keypad of olfactory receptors, each compound triggers a specific set of receptors and hence a different code.[12] Different compounds trigger different codes, which in turn yield different constellations of excited cells on the olfactory bulb. A mix of compounds, such as that found in truffles, strawberries, or cooked bacon, yields a composite constellation. But much as the Wright brothers didn't first test their plane in a hurricane, neuroscientists haven't begun with such complex mixes. What we know most about are individual compounds in isolation. But even just thinking about individual compounds, we still need one more metaphor to describe what happens in the brain in response to signals from the nose: that of a library catalogue.

Historically, when libraries began to grow in size, they needed systems of organizing the books. Different systems were devised, but the most popular systems organized the books on the shelves according to their subjects. A set of cards could then be consulted to search, within a subject, for a particular title. The larger libraries became, the more subjects had to be subdivided. The subject "herbs" came to have a subcategory for mint, which in turn had subcategories for spearmint, field mint, and French mint. Something similar happens in our brains with aromas. As we learn aromas, each becomes a subject in the mind's card catalogue, and the memories associated with that aroma are the books within that subject. But just as different libraries might use different subjects to categorize books, two people can use different subjects to categorize the same smell. Recently, Rob brought a particularly stinky washed-rind buffalo milk cheese into his class for students to smell and taste. One student, Zachary Ang, said the cheese smelled like a petting zoo. The cheese's aroma fell under the subject "petting zoo," of which it was a special case. To another student, Nathalie Mea, the cheese smelled like Cheez-It crackers. It was a special case of "Cheez-It cracker." But were those two students to continue to smell cheeses similar to the one Rob brought in, they might develop a new subject for "stinky cheeses." The brain's magical library can wheel in new shelves for new subjects when the need arises.[13] Generally speaking, the more often you smell related aromas or taste the flavors of which they are a part, the more "books" about those aromas you have, and the more finely divided the subjects in your neurological card catalogue become. Expert wine sniffers become expert in part by training their libraries and by adding to them through practice. As they do, their minds create ever finer categories for wine aromas and flavors, though never more categories than there exist olfactory receptor codes.

Each species uses its own subjects in the library of its mind, but so too each individual within each species. In this way, the libraries of minds are private rather than public. Our categorization of the olfactory world is individual. Wine experts can be similar in their ability to identify wines and yet, as Gordon Shepherd points out in another of his books, *Neuroenology*,[75] show very little overlap in how they categorize and describe the wines they have identified.[14] But there is something else individualized about our smell libraries. The subjects become ranked, good to bad. As categories of aromas populate our minds, each is attached to a set of experiences: to memories. The catalogue subject "mint" is, over years, populated with memories of the experience of smelling mint. Each of those memories consists of the memory itself, but also the emotional experiences associated with the memory. Your brain contains a ranking of the pleasantness of each aroma you have ever smelled, a ranking weighted by pleasing and displeasing memories. This ranking might be similar to those of other people with whom you have shared experiences, but never identical. It is like your brain has its own Yelp rating system in which the comments attached to each rating of each aroma are the those written through your own experiences.

Now let's return to the human evolutionary story. If *Homo erectus* and other ancient humans used fire, their olfactory libraries might have helped them learn to love cooked meats and roots. But regardless of whether or not *Homo erectus* used fire, its olfactory library must have played an important role in another context. *Homo erectus* traveled. As it traveled, it encountered new habitats. Its olfactory library allowed it to attach meaning to particular habitats. Swamps could smell like danger, forests like joy. Or the other way around. We don't know. But then within each place, each swamp, forest, or steppe, the individual fruits, seeds, roots, and other foods could also be learned.

Different flavors would be learned and loved in different years, decades, centuries, or millennia. We have an inkling of how slowly or quickly this might have happened from another set of observations made by researchers studying chimpanzees. The research group was led by Yukio Takahata; Takahata studies chimpanzees at Mahale, the same site in Tanzania where Toshisada Nishida long worked.

The chimpanzees at Mahale were originally provided with some domesticated fruits by Nishida to help to habituate them. But by 1975 that feeding had stopped (except for the occasional piece of sugarcane). Domesticated fruits were still available; they just weren't provisioned. In 1974, a government policy shift led the villages and scattered houses near to the chimpanzee habitat to be abandoned. This meant that the fruiting plants were also abandoned. Those plants, mostly trees, included bananas, guavas, oil palms, oranges, papayas, and pineapples. Suddenly, they were more accessible for the chimpanzees; they were no longer defended by old women wielding brooms and children shouting admonishments and curse words. The chimpanzees, in response, immediately began to eat the bananas. This was perhaps not a surprise; it was bananas with which Nishida had originally provisioned them when he began his work. The older chimpanzees were used to bananas. The other fruits would take more time. It was not until 1981 that the first chimpanzee was observed to come and try a guava. In the years that followed, that chimpanzee, which had learned to like guavas, continued to eat guavas, as did five other chimpanzees. But most chimpanzees never did; they did not even try them. As for mangos, a five-year-old male chimpanzee tried some unripe mangos, followed by his older brother and then a few other chimpanzees, but mango eating never caught on.[76]

Then there was the lemon tree. On June 28, 1982, an unidentified female chimpanzee from one group of chimpanzees at Mahale climbed into a lemon tree and tried a lemon. Then, in July, another adult female did the same. Finally, on August 10 an adult male ate a lemon. The next day he ate one again, and other chimpanzees gathered around him and began to eat lemons too. Within a month, twenty chimpanzees were regularly eating lemons, within a year, forty chimpanzees. In the subsequent years, the lemon tree stayed popular. The chimpanzees would break the lemons in half with their teeth and while holding one half in their feet, hold the other half up to their mouths and extract the sweet-sour contents.[77] To paraphrase the poet William Carlos Williams, the lemons tasted good to them. It was clear by the way they gave themselves to the half-sucked-out halves in their hands.

When confronted with the lemon tree, the chimpanzees were clever enough to learn to distinguish the lemon tree from other trees and the lemon fruit from other fruits. But they also came to learn to enjoy the aromas of lemons, aromas associated with lemon flavors but also with being with their community in a big lemon tree, cracking fruits open with their hands, holding them with their feet and indulging. Ancient humans would have learned, again and again, to love new aromas and flavors, the way that the chimpanzees learned to love lemons. Sometimes those flavors were new fruits, or leaves, or insects, or even mussels. On other occasions they were the aromas and flavors associated with foods that had been hidden, but could be revealed with stick tools. Ancient humans could learn the aromas and flavors of oil nuts pounded open, or of algae, long hidden beneath the surface of the water. These were new aromas, aromas magically released from the places they had been trapped

FIGURE 3.2. The card division of the Library of Congress. Each card is
categorized according to its subject, sub-subject, and sub-sub-subject.

(in nuts, for example, or underwater) using sticks. Then, eventually, came different forms of food processing.

Once our ancestors could process foods, the world had another dimension. Ancient humans almost certainly cut and
ground foods. That cutting and grinding revealed new aromas
and flavors, but only to a limited extent. Then came fire. As
we've said, no one knows with any certainty when this first happened, when this other layer of the edible world was created.
This act of creation may well have begun by the time that Neanderthals arrived in the Dordogne region in France. During
warm periods, Neanderthals appear to have cooked. They
cooked the meat of roe deer, fallow deer, wild pigs, and red deer.
These animals, once cooked, would yield aromas and flavors as
complex and wonderful as those of fruits. These were aromas

that Neanderthals might have been predisposed to love.[15] Once modern humans evolved, they began to explore even more of their environment with their mouths and discovered even more ways of creating new aromas and flavors from the ingredients around them in the world. As they did, we hypothesize, they learned to distinguish the aromas and flavors of different cooked foods. They liked some of those flavors better than others. Such preferences would come to be of great consequence. As we explore in the next chapter, these preferences may have led to the first (but not the last) culinary extinctions.[16]

CHAPTER 4

Culinary Extinction

The beasts have memory, judgement, and all the faculties and passions of our mind, in a certain degree; but no beast is a cook.

— JAMES BOSWELL, *THE LIFE OF SAMUEL JOHNSON*

Gourmands can distinguish the flavor of the thigh on which the partridge lies down from the other.

— JEAN ANTHELME BRILLAT-SAVARIN,
THE PHYSIOLOGY OF TASTE

Recently, we visited southern Arizona, ten miles from the border with Mexico. While there, we began to ponder a unusual flavor, that of mammoth meat. The flavor of mammoth meat doesn't seem as though it is particularly relevant to daily life in Arizona or anywhere else. But it is. Mammoth meat is an emblem of the flavors we have loved to oblivion.

We were staying in Patagonia, an old mining town. Today, it is perhaps best known as the home of the writer Jim Harrison, the one-eyed author of fiction (*Legends of the Fall*, most famously) and poems, known equally for his love of words and his love of food.[1] We were in Patagonia to hike, think, eat, and explore. The region is one of the most biologically diverse in the United States. In the mountains that ring Patagonia live hundreds of species of birds, but also jaguars, black bears, and bighorn sheep.

One day, Rob decided to walk down Sonoita Creek with our son. The creek is in some places above ground, some places below. The two walked a stretch near the house of the naturalist-writer Gary Nabhan in which we were staying. They walked the dry bed, beneath which flowed the living creek. As they walked, they saw the tracks of collared peccaries, both their hoofprints and the spots where these wild pigs had sniffed, nosed, and unearthed some delicacy buried just under the surface of the dry bed. They heard the calls of Chihuahua ravens (to which one inevitably feels compelled to respond—they did), and found the musty warren of a fox. In the days to come, they would also see the tracks of coyotes, spot peccaries, and see dozens of hawks waiting for the movement of mice. But while the landscape is wild, and this wildness has been a feature of the modern writing about the region, what became most conspicuous was a kind of absence. It was an absence that became more conspicuous when our son picked up a flake knocked off a piece of stone by an artisan making a tool, maybe even a spear point. Based on where the flake was in the bank, it could have been more than 10,000 years old, a vestige of someone's attempt at a very big lunch.

Understanding this landscape requires knowing a bit of river genealogy, which body of water begets which. The rivers of this landscape connect things, as they long have. The Sonoita Creek does not flow year-round. As a result, it might better be described a "draw." The draw flows, seasonally, into the Santa Cruz

River, which flows into the Gila River. In southern Arizona, the Gila is the only major river. The Gila River flows southwest to the southwestern corner of Arizona, where it empties into the Colorado River, which now empties into the mostly dry Colorado River Delta in Mexico at the northern end of the Gulf of California. Just east of Patagonia, another draw, Curry Draw, makes its way to the Gila River via the San Pedro River. It was in Curry Draw that archaeologist Vance Haynes found something far more impressive than a stone flake.

Haynes excavated a site at which he found unusually long spear points of a type called "Clovis," after the town of Clovis, New Mexico. These Clovis points were deep in the river bank, below a black band in the cut bank (it was beneath this same band that our son found a flake). The points were accompanied by other stone tools of the kinds used by paleo-peoples around the world to butcher mammals, along with several hearths, and the tusks and bones of thirteen mammoths.[2] Vance Haynes and other archaeologists would discover Clovis spear points at five additional sites along the San Pedro, and more bones of ancient mammals, some with evidence of butchering and burn marks from being cooked. These sites were banks where ancient people gathered by the river. Collectively, they have become one of the most important lenses into paleolithic life in the Americas. They are among the best-studied sites of the Clovis people and their preferences and consequences.

The Clovis culture, most archaeologists now agree, was not the first culture of the Americas, not hardly. Archaeological sites have now been found here and there in the Americas that date to long before the first Clovis points. A new site, for example, has been discovered in Chiquihuite Cave in a highland region of central-northern Mexico that was occupied no later than 30,000 years ago. The site yielded 1900 bits and pieces of ancient

stone tools across 10,000 years of habitation.[78] Similarly ancient sites have been found elsewhere, including along the coast
of Chile. These very old sites are still too few and too poorly
connected to one another to provide a clear picture of the first
humans to arrive in the Americas. More recent pre-Clovis sites,
beginning around 15,000 years ago, are more numerous and better documented,[79] and yet still pose more mysteries than answers. It is difficult to know which route these people took into
the Americas. It is difficult to know which route they took once
they were in the Americas. What we do know is that these people
or peoples were hunters who also gathered.[80] We also know that
these peoples, while traveling, hunting, and gathering, encountered a living world unlike any that their had ancestors known
(and unlike any of their descendants would know). It was a world
wild with animals and plants but also the flavors of those animals
and plants. This was the discovery of the Americas.

By the time that the first people arrived in the Americas, the
animals of Europe and Asia had been hunted by humans, including Neanderthals, for hundreds of thousands of years, and
by ancient humans for almost a million years. The edible animals of Europe had, over these millennia, learned to fear
humans. They had also become ever rarer. The first Americans,
in contrast, confronted a fauna that was new to them, naïve to
their spears, and abundant. Brillat-Savarin said that eating a
new dish "confers more happiness on humanity than the discovery of a new star." Here then was an entire solar system of
potential dishes. In North America, the first Americans encountered three times as many species of large mammals than
are found in game parks in Africa today. And even more species
awaited discovery farther south.

For thousands of years, the pre-Clovis peoples of the Americas explored the dimensions of their new culinary solar system.

They hunted. They used a diversity of stone and bone tools. For example, a 13,800-year-old mastodon (*Mammut americanum*) was recently discovered in the bottom of a pond at the Manis Site in Washington State with a bone point thrust into one of its rib bones.[81] Then, by 13,000 years ago, as the climate was warming, the Clovis culture emerged, defined by a unique type of spear point, the Clovis point.

The people who made these Clovis points could use them to more effectively kill giant sloths (of which there were five species in North America and still others farther south), mammoths, or mastodons. With these points, the Clovis people specialized in hunting and eating big animals. Such specialization was a luxury allowed when big mammals were abundant.[82] The Clovis people hunted big mammals across a geographic area that spanned Alaska to North Carolina and south to parts of Mexico. The number of sites at which big mammals have been found with Clovis points is, as archaeologists Gary Haynes (no relation to Vance Haynes) and Jarod Hutson put it, "astonishing," given that the sites tend to be in the open, where things can wash away, and the Clovis people did not occupy the sites for long—days, not weeks or years.[83]

Once killed by Clovis points, the mega-beasts of the Americas would have provided enormous quantities of meat. Like their predecessors, the Clovis people did not exclusively eat large mammals, or even meat. At some sites, the Clovis people ate hawthorn berries, for instance.[84] They may, as one archaeologist suggests, have sat around the fire spitting the berries' seeds into the hearth (where they were found) and talking. Yet, the Clovis archaeological sites are more dominated by the bones of big mammals than nearly any other archaeological sites in the world. Neanderthals are often described as the consummate meat eaters. From evidence at some sites in Europe,

FIGURE 4.1. A sampling of Clovis spear points. Notice that although the shape of these points is relatively invariant, their sizes differ, as does the type of stone used to make them. In North Carolina, for example, where we live, nearly all of the hundreds of Clovis points that have been found were made from the stone from one face of a small mountain, a hill really, in the center of the state. The Clovis people had strong preferences, whether with regard to their meat or their tools.

Neanderthals ate more meat than did hyenas living at the same sites.[85] And yet, more plant materials have been documented in the diets of Neanderthals than of Clovis peoples, even though the Neanderthals lived tens of thousands of years earlier such that the plant remains had many more opportunities to break down and disappear. The Clovis peoples ate large quantities of meat.

The Clovis people cooked much of the meat they ate. They undoubtedly did so with great skill. By the time of the occupation of the sites near the San Pedro River, humans had been cooking regularly for no fewer than a hundred thousand years, and potentially much longer.[86] That is a lot of time to practice eating and cooking. A lot of trial and error, of finding the perfect-sized fire, the perfect stick on which to hang meat, and the perfect length of time to cook the meat.

Each form of cooking, whether by the Clovis peoples, their ancestors, or contemporaries, would have entailed expertise. These people were able to make tools that involved many steps. They built homes and worked hides. They knew how to haft spear points onto spears and build atlatls with which to throw spears. They talked to each other and learned from one another. In cooking, as in tool making, they must have done their work with care. They would have had favored flavors as well as favored approaches to creating those flavors, recipes handed down from generation to generation, perhaps not anything so complex as the eight days of preparation for an authentic French cassoulet, but almost certainly more complex than we tend to envision when confronted with little more than bones and stone tools. In the *Iliad* (700 BCE), Homer has Greek priests sacrifice cattle to Apollo. When they do, they cook the sacrificial meat in one of the ways that we might imagine the Clovis people cooking, for example, a bison. They

> skinned the [cattle] and carved away the meat from the thighbones and wrapped them in fat. . . . And burned these over dried, split wood and over the quarters poured out glistening wine while young men [. . .] held five-pronged forks. Once they had burned the bones and tasted the organs they cut the rest into pieces, pierced them with spits, roasted them to a turn and pulled them off the fire.[3]

We don't know if the Clovis had wine (at some sites they lived among grapes), but otherwise this is a scene one might have encountered twelve thousand years ago in southwestern North America, or at least one of the scenes.[4] Other approaches to cooking might also have been employed. The descendants of the Clovis eventually came to use earthen ovens to slow-cook food, hot stones to boil food (in holes in the ground), and mixtures of earthen ovens and hot stones to steam food (similar approaches were used by the people of northern France some thirty thousand years ago).[87] But so far there is no evidence that the Clovis themselves baked, boiled, or steamed. Nor, for that matter, does the evidence suggest that the Clovis people used every last bit of the animals they killed (in the manner of the Neanderthals in Europe tens of thousands of years prior). For the most part, they did not seem to break open bones for marrow or burn the bones. Nor did they consume the meat from all of each individual animal they killed; theirs was a land of flavor and relative plenty.

One can't help but wonder what exactly the flavors of the ancient meats eaten by Clovis peoples might have been. We know something about the menu. It definitely included mammoths, mastodons, gomphotheres, bison, and giant horses. It might also have included Jefferson's ground sloth, giant camels, dire wolves, short-faced bears, flat-headed peccaries, long-headed peccaries, tapirs, giant llamas, giant bison,[5] stag moose, shrub-ox, and Harlan's muskox, the bones of many which have been found in or near Clovis sites. The flavor of these meats is of interest as a dinner table conversation topic, but there is also something more. The arrival of the Clovis people in Arizona and nearly everywhere else in North and Central America coincided with, or just slightly preceded, the extinction of many of the species the Clovis peoples ate. Food writers sometimes discuss extinct flavors. They write achingly of never being able to experience the extinct roman herb laser or particular kinds of asparagus. But this is different. The Clovis menu, if written on a chalkboard, would be a tally of a lost world.[6]

In the 1960s, as more and more archaeological sites were being discovered that, like Murray Spring, contained Clovis points, the bones of large mammals, and bones that showed signs of butchering, it didn't take long for someone to connect the dots.

In 1967, Paul S. Martin proposed that the Clovis prey species were extinct because the Clovis used effective hunting tools to kill and eat relatively naïve prey.[88] Martin, a geologist by training, had by then spent decades working at The Desert Laboratory (then the Carnegie Desert Botanical Laboratory), not far north of Patagonia, Arizona. While there, he considered the changes in the species found in the southwestern United States over the last twenty thousand or so years. He knew well which species had gone extinct and which had survived. The survivors, Martin argued, were either smaller species (the rats and racoons of the world) or populations of larger species that had some other way of holding out. Ground sloths were present in Cuba, Hispaniola, and Puerto Rico until humans arrived on these islands eight thousand years ago. They lasted even longer on the smaller islands that were colonized by humans later. Meanwhile, mammoths lived on until 2000 BCE on Wrangel Island in the Chukchi Sea between Russia and Alaska. They lived on, tellingly, despite the changing climate, on an island where humans were absent. Then, humans arrived and the Wrangel Island mammoths also disappeared.

The extinctions of the North American megafauna were, to varying extents, what the author Lenore Newman in her book *Lost Feast* has called culinary extinctions, extinctions brought on, at least in part, by the food preferences of humans.[89] They were not to be the last culinary extinctions. As humans arrived on islands around the world, they quickly ate into oblivion the largest species on those islands. Upon arriving in New Zealand, humans encountered eleven species of giant flightless moas. These birds appear to have been delicious. They were all quickly eaten into extinction. The dodo, it is said, tasted fine, fatty, and rich. It was not as good as pigeons or parrots, but good enough, and abundant, until it wasn't.[90] The Mauritius red rail, a flightless and defenseless bird, tasted like roast pig.[91] Humans continue to hunt delicious species into rarity. Then, once they are rare, they hunt them even more because their rarity makes them even more valuable and their taste even more special (as is currently the case for some sturgeon).[92]

Nor are the culinary extinctions of the Americas likely to have been the first. Many of the biggest beasts of Europe had already become rare or extinct in others (for example, woolly rhinos, woolly mammoths, Irish elk, and cave bears) by the time the first Clovis point was being made. Culinary extinction and endangerment are not even exclusively caused by humans. One chimpanzee community at the Ngogo research site in Kibale National Park in Uganda has developed a great fondness for hunting and eating red colobus monkeys. A recent study concluded that as a result, the monkeys are now rare where the chimpanzees are most dense.[93]

In the wake of the megafaunal extinctions in North America, ecosystems changed. With fewer species to nibble back small trees, grasslands became forests. Fires became more frequent.[94] The Clovis people also changed. Populations became more isolated. Weapons became smaller, more intricate, and more different from one place to the next, reflecting differences in diets. In some places, rabbits took the place of mammoths, in others turtles or birds. The Clovis point eventually disappeared, along with the megafauna species into which it had plunged for hundreds of years.

Hundreds and perhaps even thousands of scientific papers have, by now, been written about the role of humans in the extinction of the big mammals of the Americas and elsewhere in the world. The grumbling consensus appears to be (though some will still grumble at this statement) that the megafauna went extinct due to a mix of overhunting and changes in climate. For some species, climate change may have been the main or even the only driver.[7] For other species, hunting was the main driver. For most, some mix.[8] Resolutions of major scientific questions that, like this one, are conditional (sometimes this, sometimes that) are invariably the most hard fought. Humans like all-or-nothing answers; in this, scientists are no different from anyone else. But the ecological world and the ancient human world are rarely all or nothing. To the extent that any part of the Clovis story is reasonably all or nothing, it is with regard to what the Clovis ate. Even once the biggest mammal species started to become rare, the Clovis people appear to have hunted them, disproportionately. Even if the Clovis preference for hunting large species was only part of the reason for the demise of the megafauna, it had a significant effect.

Ecologists like to use the concept called "optimal foraging" to explain the choices that predators, hunters, and foragers make. Optimal foraging is a "best bang for your buck" view of the world, where the bang is always caloric. Optimal foraging predicts that hunters will try to maximize the number of calories they obtain in a day and so tend to spend their time hunting and foraging food items that provide the most calories for the least work. But such models assume that people (and other animals) are entirely rational, have the perfect data on the calorie content of different prey species, and care only about calories. None of these things is true, especially when it comes to hunting. For example, in many cultures male hunters spend more calories trying to kill prey items than is optimal. In addition, those same men tend to be more likely to hunt in seasons in which gathered roots, fruits, berries, or honey abound. That is to say, they hunt more when the meat they obtain is less necessary. In such contexts, some anthropologists have concluded, hunting by macho men is often as much an opportunity for men to show off as it is about optimizing caloric intake.[95] But showing off isn't the only reason that hunters might hunt in ways that aren't calorically optimal. What if the flavors of some animals, once cooked, are delicious, and those of others are revolting?

While in Patagonia, Arizona, visiting the Clovis sites (and eating at restaurants featuring the small mammals that became more common in the absence of the megafauna), we began to

MAMMOTH

FIGURE 4.2. The cuts of mammoth meat, as imagined by Ben Kaiel.

ponder the flavors of the extinct Clovis menu. We began to think about paleo-pleasures. We began to search for studies in which scientists, anthropologists, or anyone else had examined the preferences of modern hunter-gatherers for the flavors of different kinds of meat. We were hoping that the preferences documented in such studies might relate to aspects of ancient flavor preferences. Such studies basically do not exist, with one exception, a single paper by Jeremy Koster.

In 2004, during his PhD research, Koster traveled to the eastern coast of Nicaragua to work in Arang Dak and Suma Pipi, two communities in which the indigenous Mayangna and Miskito peoples live together. The Mayangna and Miskito speak related languages and are both descendants of the peoples who, by 2000 BCE, inhabited most of what is now Nicaragua.[96] There is no reason to believe that the decisions of Mayangna and Miskito peoples are culturally continuous with those of their ancestors four thousand years ago, much less those of their ancestors living in the time of Clovis people to the north. But like ancient hunters, the Mayangna and Miskito must make decisions about which species to pursue, kill and eat. Koster was interested in those decisions. Studies of hunters in the eastern forests of Nicaragua, like those of modern hunter-gatherers and the Clovis people, had basically all assumed they foraged "optimally." But these "optimal foraging" approaches didn't explain a number of things that Koster had seen on earlier visits to the two communities. For example, the hunters seemed to sometimes ignore relatively easy-to-kill prey, even big prey. Giant anteaters are enormous and relatively easy to kill but are almost never eaten. Koster thought that the decisions that traditional hunters make might be more complex than optimal foraging models would imply. He wondered if the hunters worked less hard to kill animals with flavors they did not like. Hunter-gatherers, Koster thought, were like all the rest of us. They chose what to eat based on a mix of what was available and easy to kill *and* what tasted good, or at least didn't taste bad.[9] This might sound like a minor distinction, or worse, an obvious conclusion. Yet, it was an idea to which other people studying the choices of hunters hadn't paid any attention. So, as part of a broader project, Koster decided to have these peoples describe the flavors of each of the many different animals they eat.[97]

Koster spent a year following the hunters of Arang Dak and Suma Pipi and interviewing those hunters and their families about their experience of the meat of the animals they killed

FIGURE 4.3. A Miskito woman preparing a paca, one of the favorite dishes in the region. The paca is being cooked simply, fur on, over sticks. This is one of the ways that mammals have been cooked for at least tens of thousands of years and perhaps much, much longer.

and ate. During this time, he generated a ranking of the relative tastiness and ease of cooking of each common bird or mammal species, something akin to Yelp ratings, but for the forest (see figure 4.4 for the scores). He also calculated how hard it was to find and kill each mammal and also the more commonly killed birds. If the people Koster interviewed and followed were optimally foraging with regard to the calories their prey provided and the energy expended in getting those calories, animals should be killed and eaten first if they are easy to find and kill, easy to process, and provide many calories. To some extent, this appeared to be the case. Big animals that are easy to find and kill were more likely to be eaten than animals that were smaller or harder to find and kill. This is optimal foraging, or at least one version of it. But optimal foraging didn't explain all of the decisions the hunters made.

Hunters went out of their way to kill some animals that they didn't like, such as pumas and ocelots, species they think of as their competitors. When they killed these cats, they didn't always eat them.[10] In the context of Martin's overhunting hypothesis, if Clovis people also killed predators (and there is some evidence of killing of saber-toothed tigers and wolves),[98] even if they weren't going to eat them, it might help explain the speed of the extinction of large carnivores. But there was more.

Koster observed that hunters sought every opportunity to pursue the animals they said had the best flavors. White-lipped peccaries, collared peccaries (the same species we saw in Patagonia, Arizona), and pacas were among the favored meats. These animals were pursued a hundred percent of the time they were spotted. Eastern Nicaragua, as it turns out, is a hard place to be a peccary or a paca. This might have been predicted given the ease of hunting these animals and the number of calories they provide (in essence, given optimal foraging), but anecdotally it seemed to Koster that there was an added eagerness to the pursuit of these very delicious species. In contrast, Koster observed that the hunters did not always pursue tapirs, even when they had found a tapir resting spot. Tapirs, although easy to kill and full of calories, are ranked as only moderately flavorful (Koster says they taste like chalk).

Koster couldn't demonstrate unequivocally that the hunters pursued the very tastiest animals more than optimal foraging models would suggest (whether to his own satisfaction or that of his colleagues), but it seemed to be the case. What Koster could show was that some animals that were not tasty were avoided. Howler monkeys, which are relatively easy to kill and common, were pursued just ten percent of the time they were spotted. The meat of howler monkeys is disliked.

Overall food preferences were remarkably consistent among the individuals Koster interviewed, and also consistent with regard to the consequences: most of the animals that the indigenous hunters in the communities of Arang Dak and Suma Pipi find to be flavorful tend to be rare in regions in which there is a lot of hunting.[11] Conversely, the howler monkeys abound even near settlements. But *why* are some animals more flavorful than are others, even once cooked? What makes peccaries so very tasty and howler monkey so very not? The first thing to note is that the answer, in all its complexity, will depend upon the species that is doing the eating. Jaguars, mountain lions, the extinct American lion (about twice the size of a mountain lion), and house cats alike lack sweet taste receptors, as noted in chapter 1. As a result, the sweetness of a meat is almost certainly irrelevant to their preferences for that meat. Similarly, bitter taste receptors differ greatly among species of mammals, so whether or not a particular prey item tastes bitter depends upon the compounds to which the taste receptors of a particular carnivore or omnivore species responds. Many of the bitter taste receptors in cats, to stick with the same example, are broken (see chapter 1) and so meat that might be bitter to a human might not be bitter to a cat. One might imagine that a predatory cat, whose instinct is to kill, might not care very much about taste or, more generally, flavor. And yet, among the very few fruits that cats eat are avocados, which, with their umami-rich tastes and fatty mouthfeels, approximate something meaty. It is perhaps a result of their preference for such flavors that the densities of predatory cats tend to be very high in avocado orchards, where the cats are drawn to a kind of "meat" they needn't pursue.[99] We might imagine that each predatory mammal has, in its mind, a ranking of preferred prey species, a ranking informed by the ease of killing them (optimal foraging strategy) but also by

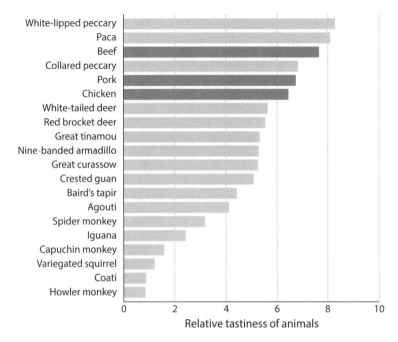

FIGURE 4.4. The relative deliciousness of different animals according to Miskito and Mayangna hunters and their families in Nicaragua. Species are ranked from the least tasty (howler monkeys) to the most (white-lipped peccaries). Species in black are domesticated animals not native to the Americas. Note that some vertebrates from the region are not regarded by the Miskito and Mayangna as food, even if relatively common, and so not included here: for example, cats and vultures.

flavor. But our focus here is on the flavors of different prey species to humans.

Some of the flavor in cooked meat comes from the protein in the muscles. Muscle flavor derives from the mouthfeel of the muscle combined with the aromas from sulfur compounds in the protein. Unfamiliar meat often "tastes like chicken," in part because the dominant flavors in chicken are these simple, somewhat bland, muscle flavors. "Tastes like chicken" really means "tastes like muscle." Muscle flavor is easy to complement with sauces or herbs, or bread batter and oil, but it is relatively non-descript on its own.[12]

More unique, variable, and subtle flavors and textures come from the fat and the collagen that weave among the muscle fibers as well as compounds that can become embedded in the fat (and to a lesser extent in the muscle and collagen). The differences among mammal and bird species in Koster's study are in part a function of differences in their amount of fat and the chemicals held in that fat. In nature, the amount of fat in an animal relates to both the habitat in which it evolved and how it lives. Plants tend to store energy in carbohydrates, but carbohydrates are not very dense. For plants, this doesn't matter because they don't need to move.[13] Animals, on the other hand, store energy as fat, which is twice as dense as carbohydrates (and hence high in calories). In colder climates, animals tend to store more fat in order to get through the winter.[100] All things equal, animals also tend to store more fat if they live and dive in water than if they live on land (think blubber). Younger animals tend to also be fatter than older ones. And animals tend to have less fat in the dry season than in the wet season in seasonal environments.[14] The variation in the amount of fat in different meats is important in your kitchen, it doesn't explain the differences Koster observed among rain forest animals, most of which tend to be on the slender side.

Fat can be consumed on its own, or can be used in cooking or fermentation. Fat contributes to flavors in several ways. Fat contributes a smooth mouthfeel to food. The mouthfeel of fat is pleasing to the tongue as it explores and works the food it has been offered. In addition, fatty acids add taste to food, though as we noted in the first chapter, this taste tends to be unpleasant. Neither the mouthfeel of fat nor the flavor of fatty acids likely explains Koster's observations. Instead, Koster argued that his observations seemed to depend on the ways in which fat traps the flavors of an animal's life. Which flavors fat traps depend greatly on the details of an animal's gut and diet.[15]

When an animal eats, some of the chemical compounds from its food enter its bloodstream: proteins, fats, and sugars, but also the myriad other compounds found in the meal. A subset of those compounds is deposited along with fat in the cells of the meat of the animal. Once there, the molecules of compounds from the food bind to the fat much in the way that the odors of blue cheese or half of an onion attach themselves to an exposed stick of butter in your refrigerator. These molecules are present in raw meat and are perceived via retronasal olfaction, in our mouths. But, as meat is cooked, these molecules mix in complex ways to form additional compounds, many of which are poorly studied and understood. The aromas from such compounds are important not only to humans, but also to other species that eat meat, including dogs.[101]

At least from the perspective of human perception, the flavors that result from the aromas that are trapped in fat appear to vary predictably among animals as a function of their lifestyles. Predatory animals tend to encounter relatively few unusual compounds in the animals they eat. They also tend to be relatively lean and so have little fat in which such compounds might become trapped. As a result, predators tend to taste like a low-fat beef round roast or something equivalent (albeit often chewy round roast), unless they have eaten something with especially strong flavors (for example, ants), in which case those flavors may still be detectable in the meat.[16]

In omnivores, such as peccaries or bears, and herbivores things get more complex. The meat of omnivores and herbivores is influenced by the flavors in the foods they eat and by how effective their guts are at processing the compounds that produce those flavors. In general, animals with guts that are very effective at digesting foods and the toxins in them tend to have meat that is not particularly flavorful and tends to vary relatively little from place to place and season to season. Their meat is the forest version of the little restaurant you like because it is always pleasant, pleasant but rarely thrilling. Animals with such dependable, predictable meat include ruminants: herbivores such as bison, cows, goats, deer, and giraffes that have rumens, stomachs with multiple chambers in which the plant material they eat is slowly and repeatedly fermented. In these ruminating animals, bacteria break down both the carbohydrates and toxins in plants into fatty acids. Those acids add subtle flavor to the fat of animals, but the effect is weak and typically hard to describe. Chefs will often mention a "grassiness" or a "vague and yet not unpleasant hint of skunk" in the meat of ruminants. Deer are ruminants. The Mayangna and Miskito ranked the two deer species they eat as tasty but not among the most delicious.

Animals with guts that are less effective at digesting food and breaking down toxins are more likely to have meat that bears the flavors of what they have eaten. The list of such animals, with less complete digestion, includes many more species. It includes species with hind guts (guts that occur after their stomach in the gastrointestinal assembly line), species with small foreguts (in which food stays too short a time to break down completely), and a variety of special cases. If animals from this second group of organisms eat foods with pleasing flavors, such as fruit and roots, their meat often strongly tastes of those flavors.[17] As the Danish ornithologist Jon Fjeldså put it, the variation in flavor of these species "reflects their diet." Such is the case for fruit-eating monkeys and wild pigs.[102] It can also true of horses. Their meat is flavored by what Guy de

Maupassant called the "quintessence of all the food that (they have) consumed."[103] They have a terroir, a flavor of the land and the time and the place in which they have lived and been killed, a flavor rich in details, history, and context.

Hunters and pastoralists alike often know when and where to hunt a species in order to encounter a longed-for set of flavors, and also how to feature these flavors of place and season. For example, Gary Nabhan told us that in Lebanon "sheep pastured in the hills in the summer taste, deliciously, of thyme and zaatar by fall. In the southwest, the Navajo prefer to eat animals that have been feeding on sage (*Artemesia tridentata*)." Meanwhile, Fjeldså writes that grouse or ptarmigan "should be left hanging outdoors for a couple of weeks before you pluck and rinse them, as the flavor from blueberries and various seeds and plant buds in their crops then spreads out through the body and gives a wonderful spicing."[18]

Most of the preferred meats of the Mayangna and Miskito are from species whose meat bears the flavor of what they have eaten and that tend to eat foods with pleasing flavors. The Mayangna and Miskito rank the two peccary species they hunt, both of which feed heavily on roots, fruits, and seeds, as the tastiest of wild mammals. So too do hunters throughout the Americas. What is more, they prefer the peccary the most when it has been eating the bulbs of particular plants, peccary with a hint of allium, for instance, or wild hyacinth. Similarly, the Hadza hunter-gatherers of Tanzania find warthogs, distant relatives of peccaries, to be delicious.[19] If the warthogs have been eating the roots of wild ginger (which they often do), the roots lend a spiciness to their meat.[20] In the 1700s, in France, it was wild boars (sangliers) that were viewed as the tastiest of animals. The taste, it was said, came from their wildness, but also their courage. The meat of a warrior pig tasted the best.[21]

Another of the omnivorous species that, for the Mayangna and Miskito, has good flavor is the paca. Pacas are cat-sized rodents that feed on roots, fruits, nuts, and the occasional insect. Pacas have short, simple guts and eat many fruits that flavor their meat and, to the Mayangna and Miskito, make it delicious. The Mayangna and Miskito aren't alone in this opinion. Charles Darwin tried either a paca or its close relative, the agouti, and regarded its meat as the best he had ever tasted (he also quite enjoyed armadillo).[104] There are no pacas in Africa, but duikers are small-bodied ruminants that eat diets similar to those of pacas and are enjoyed greatly.

As for primates, fruit-eating primates such as spider monkeys in the Americas and guenon monkeys in Africa tend to be considered the most flavorful, to the extent that they are all now rare in many countries. The naturalist Henry Bates called the meat of spider monkeys "the best flavored" he had ever tasted, resembling beef but with a sweeter and richer taste. This is a sentiment to which the Mayangna and Miskito would relate; they also find spider monkeys relatively tasty, at least compared to other monkey species.

Just as some animals pick up good flavors from their foods, others pick up bad ones. In general, the less pleasant the aromas of the foods that an herbivorous or omnivorous animal is eating, the less pleasant its meat will be. Jon Fjeldså reports that summer or fall grouse and ptarmigan have a wonderful spicing, while the same animals in winter taste of turpentine, thanks to the resinous trees and shrubs they eat during hard times. Similarly, tropical animals that eat tree leaves that are chemically well defended (as opposed to less chemically well-defended grasses) also often taste unpleasant. For the Mayangna and Miskito, tree-leaf-feeding animals such as howler monkeys are disliked.[105]22 The meat of these animals bears the tastes and flavors of the bitter compounds in the leaves they eat. It is probably not a coincidence that howler monkeys are one of the most frequently taboo species in the Americas. It is easy to have a taboo against a species no one wants to eat in the first place. The dislike of howler monkeys across the Americas is in line with the preferences of the Mayangna and Miskito, who rank the meat of howler monkeys as the very worst.

Overall, the influence of the guts and diet of animals on the flavors of their meat appear to explain most of why the Miskito and Mayangna prefer some meats over others, why they love peccaries and pacas, like spider monkeys, and dislike howler monkeys. Where data exist, preferences throughout the tropical Americas appear to be relatively similar. This is the case even for cultural groups that have been separate from the Miskito and Mayangna for thousands of years, such as the Waorani in Ecuador (figure 4.5). The rank order of mammals preferred by the Waorani is nearly identical to that for the Miskito and Mayangna. Data for other parts of the tropical world are sparse but tend to show similar patterns. The species that are delicious are reasonably predictable. With few exceptions, they also tend to be ever more rare; endangered by their deliciousness.

Given what we have learned from the consideration of modern hunters, we can revisit the story of the Clovis peoples. But before we do, let's acknowledge some limits. The peoples we know the most about with regard to preferences for particular meats are tropical hunters and hunter-gatherers. The environments inhabited by the Clovis people were varied and included temperate rain forests and temperate deciduous forests. But in much of North America, they would have been cool, grassy steppes, punctuated by clumps of trees. Unfortunately, the flavors preferred by the modern hunter-gatherers in such cooler

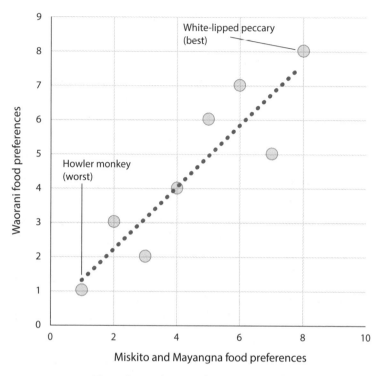

FIGURE 4.5. The preference for particular meats among the Miskito and Mayangna in Nicaragua correlated with the preferences among the Waorani people in Ecuador. The groups share little in terms of culture, language, or modern history, and yet to both groups pacas and peccaries are as good as it gets with regard to flavor, and howler monkeys are the very worst.

climates have been poorly studied. Another limit relates to how meat was prepared. It is very likely that much of the meat was cooked, over fires, in ways that emphasized the meats' own flavors. But the meat might have also been fermented or seasoned or cooked for long periods in stews. Chef Kim Wejendorp pointed out to us that while meat of older animals tends to be tougher (and so less preferred for cooking over the fire) it also tends to be more flavorful, especially in stews. As a result, if the Clovis were slow-cooking stews, they might have preferred different meat (older, say) than if they were just roasting meat over the fire. No evidence yet exists of Clovis boiling or stew-making. Nor have any Clovis grinding or food-processing stones or other technologies or vessels been found, but a great deal of food preparation can happen in vessels made of materials that don't last well in the archaeological record, such as skins. In addition, different cultures, and even different individuals within a culture, can come to prefer different meats and different ways of preparing meats. For tropical peoples, it seems as though preferences seem to be similar from one region to another, but it might not have been the same for the Clovis. Behavioral biologist (and lover of food) Carlos Martinez del Rio, for example, pointed out while reading this chapter that some ruminants, such as the pronghorn, have meat that is delicious and strongly flavored, regardless of their being ruminants. Yet he was also quick to point out that his wife, Martha, does not like the flavor of pronghorn. Finally, while many cultures prefer fat in food and the mouthfeel of fat in meat, Sophie Coe, author of *America's First Cuisines*,[106] found that pre-colonially such flavors were disliked by many Mayans and Aztecs, who were disgusted by the European use of fat. On the other hand, many peoples of the far north rely heavily on fat and even fermented fat in their diets. Mysteries remain, plenty of them.

With all of those caveats and mysteries in mind, we can first hypothesize that the Clovis people would have noticed the flavors in their foods, paid attention to those flavors, and had preferences for some flavors relative to others. Such a possibility seems obvious, but appears to have gone little mentioned. We can further hypothesize that they might have been satisfied by the meat of ruminants, such as the giant bison, with their pleasant but often bland flavors. But they are likely to have preferred meat from non-ruminant animals, especially if the diets of those animals included fruit and roots and weren't too dominated by tree leaves.²³ From the list of species present during Clovis times in Arizona, such animals would have included, to varying extents, mammoths, mastodons, and gomphotheres, all of which are non-ruminants with diets that included fruits, grasses (mammoths), and, in most places, cold-climate tree leaves (mastodons).²⁴ As mammalogist Joanna Lambert pointed out to us, the species of tree leaves in the mastodon diet would have likely been defended by tannins rather than more toxic defenses. Tannins are a kind of all-purpose plant defense. They are found in grape skins, oak leaves, and many other plant parts and species. They bind to the proteins in animal mouths, including the proteins that make saliva slippery, and trigger a sensation that is described as "astringent." Tannins make us pucker and, ever so slightly, wince. But tannins, unlike more potent plant defenses, don't tend to end up in meat. In short, the big animals that the Clovis have been documented to have killed would have all probably been pretty darn tasty. (Conversely, the largest predators and the carrion feeders were probably not.²⁵)

The Clovis did not just choose which species to eat. They also chose what body parts to eat. They ate what they most wanted and left the rest.[107] One determinant of the flavor of a particular piece of animal meat is its redness. The redness of meat depends primarily upon how much the muscle that makes up the meat moved in life and how. Some muscles are used in bursts. The quail in southern Arizona, if startled, explode like a storm out of the shrubs. They fly quickly and for a short distance. That burst requires fast twitch muscle fibers in the wings of the quail in which a sugary energy source, glycogen, is stored. Fast twitch muscles use oxygen to burn glycogen but can only do so until oxygen is depleted.²⁶ In terms of its flavor, fast twitch muscle is, all things equal, uncomplex. It is white meat, seasoned with the sugar that the muscle would have used for its next rapid movement. The leg muscles of mammoths would have been, relative to other muscles, whiter meat, more dominated by fast twitch muscle fibers. Conversely, slow twitch muscle tends to be laced with fat, fat that can slowly be converted to energy over prolonged periods of muscle use, the sort of use associated with standing and slowly walking. Slow twitch muscle is red meat. Many people in many places and times have preferred the red meat; the Clovis might have too. The reddest meat of a mammoth would have included the meat of its back (ribs) and shoulders and neck (chuck) and maybe even feet.

But an animal is more than muscle. It has been speculated that Clovis hunters may have enjoyed the organs of the animals they killed. Among many cultures, the flavor of organs is highly valued. The Clovis might also have eaten the guts of animals. Doing so would have had dietary benefits. Animals, including humans, that eat too much protein can suffer from a variety of health problems. Eating animal guts would have buffered Clovis hunters from such problems, because intestines and their partially digested contents are relatively low in protein but relatively high in vitamins, fat, and carbohydrates. At the time of the European colonization of the Americas, gut eating (gastrophagy) was common among Native Americans. It is not uncommon among hunter-gatherers and horticulturalists globally. The preparation of animal guts, and their partially digested contents, can be simple. They can be eaten raw. But preparation can also involve more complexity, various steps of cleaning, roasting, and even fermenting.[108] There is no reason to believe that gut-eating, gastrophagy, was not practiced by the Clovis; the megafauna had very big guts.

All of this together paints a picture in which majestic mammoths and mastodons, storming across the steppe, trumpeting and mating and seeking out food, animals that ate fruit and nuts and poorly defended leaves, would have had meat with lovely flavors and, at least when conditions were good, a fatty mouthfeel. They would have been delicious. This is partially supposition, but not entirely.

Woolly mammoths, mastodons, and gomphotheres were all relatives of elephants; they were all proboscideans. Other than size,[27] the differences among these species in the biology of their fat and muscles are likely to have been modest, primarily a function of differences in their diets. They would have had a background proboscidean flavor, with a hint of this or a trace of that. The mastodons of Florida, for instance, appear to have eaten cypress leaves along with various nuts and fruits. Their meat might have had an herbal, nutty flavor as a result. Mastodons elsewhere ate different plants. Mammoths tended to eat more grass. Yet, overall, the taste of extinct proboscideans is likely to have been relatively similar to that of extant ones. This is useful inasmuch as something is known about the flavors of elephants.

In a recent study of the importance of elephants to paleolithic peoples in Europe and the Middle East, Hagar Reshef and Ran Barkai at Tel Aviv University conclude that elephants are and have long been delicious.[28][109] Or, at least, some parts of elephants—whether the Asian elephant, the African savanna elephant, or the African forest elephant—can be delicious. It is now illegal to hunt and eat elephants, but it hasn't always been. Reshef and Barkai point out that both Liangula hunter-gatherers in eastern Kenya and the Nuer people of South Sudan regarded elephant as the very tastiest meat. It was, according to the Liangula and Nuer, both fat and sweet.[29]

Some elephant body parts, however, taste better than others. Mammoth biologist Gary Haynes, for example, reported that the muscle of the elephant rump is stringy, while Reshef and Barkai point out that elephant feet appear to be very satisfying. The naturalist Samuel White Baker describes the preparation of elephant feet:

> The foot of the elephant will be perfectly baked and the sole will separate like a shoe, and expose a delicate substance that, with a little oil and vinegar, together with an allowance of pepper and salt, is a delicious dish that will feed about fifty men.[30]

This sounds a little bit like the Cantonese black vinegar pig knuckles, a delicacy prepared in honor of the birth of a baby. Like the elephant foot recipe, the pig knuckle recipe includes vinegar (black), oil (sesame), and feet (the pigs' knuckles). The only added ingredients in the Chinese recipes are ginger and sugar. Perhaps the elephant feet would have been even better with this added zing and sweetness.

Samuel Baker's experience mirrors that of François Le Vaillant, who, upon eating a forest elephant with Khoisan hunter-gatherers in the 1800s, noted that he could "not conceive how an animal so heavy and coarse as the elephant could produce such tender and delicate flesh." Like Baker, Vaillant also called special attention to the feet. "I devoured," he said, "without bread my elephant's foot." The meat of the foot would have included the muscle itself, which is red and fat-riddled due to its chronic need to do work, but also the fatty pad beneath the elephant's toes that helps the animal to balance and detect vibrations.

Other humans may have also enjoyed elephant feet. At a 500,000-year-old archaeological site in Greece an elephant was found with one of its feet apparently removed using stone tools.[110] At a recently discovered Neanderthal site in Italy,

Poggetti Vecchi, bones of many individual elephants were found alongside Neanderthal tools. The tools included scrapers and digging sticks, but no spears. However, the elephants show signs of having been butchered, and the tools show signs of having been used to do butchering. In addition, intriguingly, some of the bones of the elephants were missing, as though they'd been carried off to be eaten. They were the rib bones and the bones of the feet.[111] Nor was it just Neanderthals who ate this way. The feet of one of the mammoths killed at the very first Clovis site ever discovered, in Clovis, New Mexico, were dismembered, an effort apparently directed, as the authors of that study suggest, "toward the removal of fat-rich pads from the feet."[112]

The apparent tastiness of proboscideans highlighted by Reshef and Barkai has consequences. It suggests a simple explanation for why Clovis hunters might have hunted megafauna even after they had begun to become rare, hunted them until they were gone. They were tasty enough to be worth chasing down even when the chase was suboptimal. Ancient hunter-gatherers were capable of seeking out pleasure and even going to great lengths in its pursuit. Such pleasures included art. But we hypothesize that they also included food. These hunter-gatherers had our tongues and noses, and they had our brains. They had, as a result, many of our desires.

Imagine, then, the Clovis people living in and around what are now Clovis, New Mexico, and Patagonia, Arizona, as poet gourmands. They hunted, gathered, and hiked together. They sat around fires and told stories. Some of the stories would have been powerful and moving, others would have been funny, and others still would have been boring. In the mornings and evenings, they ate gathered bits and pieces of plants and, when they were lucky, maybe mammoth foot, perhaps accompanied by wild honey and fruit. Sometimes these meals were pleasing. Other times, less so. Some cooks were good, others less so. When the food was good, it was noticed and mentioned and talked about. "Do you remember," they might even have said, "the roasted mammoth we had on the hill in the fall, when the sun was setting, the mammoth with the hackberries?" And the others would have nodded, as we do now, as we have done around the world in similar moments for hundreds of thousands of years.

In considering such moments, and the history of hunter-gatherers, food, and pleasure, we are reminded of an ancient site where hunter-gatherers once killed megafauna, the Dordogne region of France. The modern Dordogne and neighboring Cahors are home to complex wines, dozens of unique cheeses, and truffles, as well as cassoulet. But the Dordogne is also one of the most important regions in the more ancient story of humans and food. It was in the Dordogne where the bones of Neanderthals were first discovered, in a small town called Moustier. The stone tools used by Neanderthals in food preparation and hunting between 300,000 years ago and their extinction, 40,000 years ago, are now called Mousterian as a result, as is the time period. It is in the Dordogne that the bones of early modern humans were first discovered in a hole, *cros* in the local language of Occitan, in a limestone cliff owned by the Magnon family; thus these humans used to be known as Cro-Magnon.[31] And it is in the Dordogne, nearly 40,000 years ago, where those humans painted some of their greatest masterworks on the walls and ceilings of caves.

The oldest cave paintings were abstract. Moon-like circles. Rows of dots. Squares attached to lines. Hand prints. Mark Rothko and Jackson Pollock didn't invent abstract art, they revisited it. Such abstractions never disappeared from paleolithic art,[32] but did come, with time, to be accompanied by more depictive scenes, scenes of species paleo peoples ate. The depictive paintings almost never include smaller food items such as fish or rabbits. Nor do they often depict humans (when they do, the humans are very simply rendered and tend to have features that suggest magic of some sort). They never depict plants, not a single berry or herb, even though plants were being eaten. Instead, they tend to depict big prey species such as reindeer, horses, ibex, and mammoths. They depict the biggest of those species, even once those species had become rare thanks to some mix of overhunting and climate change.

Often, the big animals in cave paintings are depicted alongside babies of the same species. When we look at the babies now, we are struck by their realism. We also feel empathy (or can, anyway). The mother mammoths and baby mammoths remind us of our own families. But maybe that isn't what the artists were imagining. Consider two of the strangest cave paintings in Europe. In Rouffignac cave in the Dordogne region, a cave we recently visited, there is a painting of a baby mammoth alongside many other mammoths in a gallery a kilometer deep in the cave. The baby mammoth's feet are strangely enormous. The other painting, in Chauvet cave, is very similar, but fifteen thousand years older. It also depicts a baby mammoth with giant feet, so big as to be slightly absurd. The art in the two caves is separated by more years than separate us from the extinction of the last mammoth. Yet, the art works are similar.

The standard explanation for the size of the feet in these paintings is that the artists were trying to show not only the animal but also the bottoms of its feet and the kind of footprints it would leave. Alternatively, it is also possible that these two unusual baby mammoth paintings are just due to variation in the skill level of different paleo artists. "You know, Tom just couldn't ever paint feet, bless his heart." On the other hand, it also seems possible that these depictions of baby mammoths were pleas, prayers put forth to the gods. Oh, won't you please let me eat just one more baby mammoth with giant, delicious feet, please. The gods in the time of Chauvet cave responded. Mammoths at the time of the Chauvet cave were still commonly killed. But by the time of Rouffignac, mammoths were already nearly all gone; the plea, however earnestly painted, was almost certainly in vain.

The idea that the cave painters were depicting the mammoth feet because of their deliciousness might be a stretch. But it is not a stretch to note that the artists painting the megafauna were aware of what was and was not delicious. They knew the flavor of mammoth feet, even if it isn't why they painted them. In the end, the point about flavor and megafauna is not that flavor is the whole story of which species early Americans ate or when and why they went extinct. Cultural taboos mattered, as did cultural preferences, as did the ease of hunting one species relative to another. We know this. Instead, our point is that in nearly every discussion of choices hunters in general and hunter-gatherers in particular have made and do make, whether the Miskito peoples in Nicaragua, the Clovis hunters in ancient North America, or Neanderthals, we have ignored flavor. By considering flavor, our perspective on the choices people made and make changes.[33] The preferences of ancient hunter-gatherers matter in thinking about the story of the mammoth. And the preferences of the mammoth matter in thinking about fruit.[34]

CHAPTER 5

Forbidden Fruits

Since Eve ate apples, much depends on dinner.

—LORD BYRON, *DON JUAN*

The apple doesn't fall far from the tree.

From a culinary perspective, the extinction of the megafauna is primarily notable as an absence. But not entirely. The legacies of the megafauna also appear, rather regularly, in fruit salads. When you eat a mango or even a pear, you are biting into a complex story of extinction and flavor. It is a story about the world's fattiest, sweetest, biggest, and most aromatic fruits. It is a story first about the flavors the megafauna loved and then, also, in the end, about the ones we do too.

The goodness of fruits is buried in our language. If something pays off, it is fruitful. If it pays off without much investment, it is a windfall (of fruit). If it doesn't pay off, it is fruitless. If something wonderful is easy to obtain, its fruit is low-hanging. "Paradise," meanwhile, is a word that comes from the Persian for a walled enclosure, an enclosure that in Hebrew came to be synonymous with orchard. Paradise is an orchard.[1] The fruits of trees in orchards and wilder places evolved for just one reason, as a way for plants to attract animals to gather and carry their seeds to what might be a better place. The apple was never tempting Eve to sin. It was instead, like the mango, the peach, and the guava, tempting her to carry its seeds and then (in vacating her bowels later) disperse them. In giving in to temptation, Eve shat and fertilized a baby tree.

It might seem silly for fruiting trees to depend on the guts of animals for their success; it isn't. Seeds carried by animals can travel to new regions and habitats. On the other hand, seeds that fall beneath their mother plant are shaded as they grow. Their roots struggle to find leftovers in the soil, whatever nutrients might have been missed by their mother's bigger roots. Seedlings that grow beneath their mothers are also more likely to suffer their mothers' pathogens and pests. By endowing seeds with fruits, mother plants help to keep themselves from killing their own babies.[2]

The trick for trees and other plants that would enlist animals as transport is that they must first cater to specific animals, wooing them from afar. "I am here." Then, once those animals are close, they must beckon even more conspicuously. "Eat me." Then, finally, once the animal has bitten in, wrapped its mouth around the body of the fruit, the fruit must be good enough that the animal chooses to swallow it.[113] In general, plants have evolved the ability to participate in this dance by being attractively colored at a distance, attractively scented up close, and then flavorful. But such attraction is costly. Plants must spare their own carbohydrates, fat, and protein in order to offer rewards; they give of themselves on behalf of their children. As a result, they make fruits that are only as flavorful as they need to be and not one bit more. A plant will trick an animal, if it can, as shown with the example of the African shrub, *Pentadiplandra brazzeana,* in chapter 2, that offers little more than pleasure to those that eat it.

Fruits are among the few things in nature that have evolved to be delicious and enticing to other species. Of course, just how you entice an animal to bite depends on the sort of animal you are trying to attract. The bait must suit the quarry. Birds tend to be most easily wooed by colorful fruits that contrast with their backgrounds and so are easier for them to see.[114] Such fruits are often red, as with holly berries, cherries, red currants, or rosehips, though they can also be blue. Mammals care less about color (many fruits that attract them are green or even brown) but enjoy fatty fruits with a "fruity" aroma. At least in some regions, mammal-dispersed fruits increase their production of aroma compounds when they are ripe and ready to be eaten. Ripe bananas and peaches say, "Come to me." Bats like the smells of terpenes and sulfurous compounds, which make the fruits easier to find in the dark, but care even less about fruit color than do other mammals. Ants prefer fruits that are tiny, fatty, and, sometimes, bear the heady and peculiar scent of rotting meat. These are tendencies more than universal rules, but nonetheless you can tell a lot about the wild ecology of a fruit in the store by looking at it and sniffing it. You can tell even more once you take a bite.[115] Some ecologists who study fruit and seeds categorize the basic ways in which fruits attract animals as dispersal "syndromes," wherein a syndrome describes a set of attributes that fruits with a particular mode of dispersal tend to have.[3]

Daniel Janzen was thinking about the ways in which plants woo animals with their fruits when he came upon a mystery. He discovered a tree species with big, aromatic fruits. The fruits conspicuously advertised themselves. Yet, even when ripe, they tended to hang on their trees, unpicked. Theirs was a reward no one wanted. A key part of the story was missing: the disperser.

Recently, we visited the house where Dan and his wife, Winnie Halwachs, live for half of every year (when they are not in Philadelphia).[4] The house is in a tropical dry forest in Costa Rica. We met Winnie at the front door and she said Dan was out back, and so we walked down the trail. When we found him, he was shirtless. His leathered skin was speckled with strands of long, simian fur. His head was ringed by a tufty white halo of hair. As we watched, butterflies circled him. He looked as if he might have been created right there by some jungle god out of a mix of clay and leaves. He held a bag containing specimens of some sort in one hand, and gesticulated with the other. He gesticulated as he explained the forest around him, one sweeping idea at a time. It was the sort of hand waving that has kept him busy throughout his career, the sort of hand waving that has also reshaped how we understand the world in general and the world of fruits and their flavors in particular.

Over decades, Janzen, who was 79 when we visited with him, has written scientific papers arguing, variously, that the smell of rot is produced by bacteria that are trying to keep mammals away from the dead bodies they would rather have to themselves,[116] that rain forest trees have evolved hollowness in order to allow bats to colonize and then fertilize them with their falling feces,[117] and that tropical forests are diverse because herbivores and parasites prevent plants from growing next to their mothers and hence ever establishing dominance.[118] But his idea with the biggest effect on how we think about food was the one about the uneaten fruits.

The forest in which Janzen lives is full aromas, tastes, sounds, and feels. It is a forest in which everything eats and everything dies, but no two species eat or die in exactly the same way; a forest of capuchin monkeys and howler monkeys, spider monkeys and spiders and hundreds of thousands, perhaps millions, of other species. It is a forest in which one might be forgiven for overlooking details. Janzen, however, does not overlook details. He gathers them and leverages them as a way to understand generalities. He travels from the specific to the universal. The detail that got Janzen thinking about fruit was the unpicked fruits of a tree species called *Cassia grandis*, the "stinking toe" tree. Each fruit of the stinking toe tree is about half a meter long, hard as a rock, and shaped like a stockinged leg and foot. A tree might contain hundreds of such leglike fruits, each of which is filled with seeds the shape (and about half the size) of checker pieces. The seeds are covered in sticky, sweet flesh. The flesh has the consistency of molasses and is edible. It has a strong aroma. It also has a strong flavor that, to chef Andrew Zimmerman, is like "anchovies and fish sauce mixed with molasses," in a good way.

The stinking toe tree spends enormous energy on its strange, stinking, delicious (to some) fruits; yet it was these same fruits that were not being eaten or dispersed.[119] Instead, they hung on the tree for weeks or even months, and while they did, fungi climbed through the pods. Then the fruits fell to the ground, where beetles drilled holes through the seeds. Ants and rodents picked up the pieces and carried them away in every direction. As a result, few seedlings of these *Cassia* trees grew beneath their mothers.

As Janzen walked the forest, he noticed this tree species was not alone in its behavior. The seeds of the wild relatives of avocados were also dying beneath their mother trees, inside the rotting pulp of their fruits. And it wasn't just wild avocados and stinking toe fruits. Wild papayas, mesquite pods, sapodillas, custard apples, lucumas (kin to the cashew), jicaro, hog plum, and squashes, as well as many other similarly large fruits so unusual as to not yet have English names, were left to be devoured by some mix of insects, fungi, and bacteria.

Each of these undispersed fruits had its own aroma, taste, and shape, its precise and unique biochemistry of attraction. Some, like mangos and avocados, had enormous seeds at their centers, hard to bite and often toxic. Others, like papayas, had many tiny, soft (and sometimes slippery) seeds. Some had fatty pulp, others had sweet pulp, still others had fleshy and somewhat bland pulp. In short, the fruits varied. And yet nearly all were big, indehiscent (they didn't break open and release their seeds on their own), and aromatic. These fruits seemed to imply big, missing dispersers. Tellingly, the fruits looked to Janzen liked the sorts of fruits eaten in Africa by elephants and other large mammals.

A few years before Janzen began to notice the dangling, dying, stinking toe fruits, Paul Martin published his radically new history of the Earth's wild places. Martin had begun to advocate his specific hypothesis that many of the biggest mammals in the Americas had gone extinct when the Clovis people and their descendants arrived and began to hunt them to extinction. Like Janzen, Martin loved big ideas, ideas full of grandeur and elegance and yet contingent on details. Just like Janzen, Martin based many of his ideas on careful observations of phenomena others had taken for granted. The two scientists were a good match in general, but especially given the specific idea Janzen was pondering.

Janzen had begun to think that the reason the stinking toe fruits were left dying on trees was that the mouths and guts of the big mammals were missing. Perhaps, he hypothesized, trees

FIGURE 5.1. Fruits of the tree *Cassia grandis*, the stinking toe tree. It is unclear whether this common name, used in Belize, corresponds to the shape of the fruit or the aroma of the flesh surrounding its seeds.

with big fruits were not getting dispersed in Costa Rica because the mammal species that used to eat those fruits, carry them to new places in their bellies and, upon vacating their bowels, disperse them, were gone. Extinct. The big, delicious mammals were gone, and in their absence hung the big, sometimes delicious, fruits.

This would explain why the unpicked fruits tended to be big: they'd evolved to attract big mammals. It explained why the husks of so many of the fruits were hard to break through: they'd evolved to prevent smaller mammals from getting in. It also explained why many of the seeds were big and tough: they had evolved to be as big as possible to ensure germination once they passed through a mammal, but tough enough to escape the big teeth of the big mammals. It even explained why some of the seeds that weren't tough were small and squishy: they'd evolved to escape big teeth by slipping free from them or sliding between them. Meanwhile, the differences among these big fruits might correspond to differences among big extinct mammals and the details of their guts, noses, and taste preferences.

Janzen had one problem, though, in developing this idea. He didn't know as much as he wanted to know about extinct mammals, particularly those of Central America. In October of 1977, he sent Paul Martin a letter inviting him to write a paper with him about big fruits and mega-mammals with mega-mouths. It began, "I've got a screwy idea. . . . Let's write a paper together." Martin agreed to the plan, responding "what a fun letter and idea. I'll invoke the ghosts of some hungry extinct herbivores and you will see if they eat up that fallen fruit." Which is more or less what happened. Martin helped Janzen flesh out just which species of megafauna had been lost in Costa Rica and what they might have eaten.

Martin told Janzen that as recently as seven thousand years ago, the fruit-eating mega-beasts of Costa Rica would have included several species of giant ground sloths, which could have eaten big fruits while they were still on the trees. They included smaller bear-sized ground sloths that could have eaten fruit that fell onto the ground. They also included a variety of species of the elephant order (Proboscidea) such as species of mastodons and gomphotheres,[120] that could have eaten fruit any and everywhere. Bear-sized armadillos called glyptodonts, giant peccaries, giant tortoises, and tropical horses were were included in the menagerie. If any of these animals ate the stinking toe fruit, Martin noted, especially the mastodons or gomphotheres, the seeds would find themselves "in a fine big turd of . . . manure" in which to grow.[121]

Together, using Janzen's idea and Martin's knowledge of ancient mammals, the two scientists wrote a paper. In 1982, four years after the collaboration began, the paper was published in the journal *Science* with the title "Neotropical anachronisms: The fruit the gomphotheres ate." The paper was at once sad, wild, and wonderful. Its central idea was unusual, almost, as the journal's editor said, "like a movie script."

If Janzen and Martin were right, their idea had practical implications. If the absence of the megafauna was leading big-fruited trees to fail to be dispersed, perhaps adding back megafauna would help to disperse those trees. What Janzen needed were some megafauna. He couldn't clone the extinct species. But he could employ the relatives of those species, at least one of those species. Wild horses once roamed Costa Rica. Janzen thought that although those extinct wild horses were only distantly related to the horses domesticated in Eurasia and brought to the Americas by the Spanish, their tongues, noses, mouths, and guts might be similar enough to those of the extinct horses to serve as a kind of simulacrum, a bodily stand-in.[5]

Janzen presented horses with the megafauna fruits of the jicaro tree, *Crescentia alata*. The jicaro tree has fruits the size and shape of large oranges. Jicaro fruits have long been cut open, initially with stone tools and later with machetes, by the indigenous peoples of Central America to make bowls. Even with such tools, they are hard to split. Janzen gave the fruits to horses. If the horses could break open the fruits, they might both eat and, in their feces, disperse them.[122]

Horses are able to generate about 550 kilograms of pressure with their mouths (by comparison, the pressure of a human bite maxes out at about 70 kilograms of pressure, and even that is achieved only with the molars). In Janzen's experiments, that was enough to break open most of the jicaro fruits in order to eat the black pulp surrounding their seeds. But some of the jicaro fruits were too tough for the horses to crack. The fruits had evolved to be dispersed by a mammal with a mouth as strong, if not stronger, than that of the horses. The toughness of the jicaro fruit is an adaptation that keeps its seeds relatively safe from anything but the very biggest dispersers. Their pulp is nearly totally unavailable to any mammal smaller than a horse.

After eating the softer of the jicaro fruits, the horses walked around the field, depositing seeds. The seeds germinated in amongst the nutrient-rich feces of the horses. In their wake, a forest grew. It wasn't quite so simple, but nearly so. Wild animals now walk beneath the canopy of trees planted by domestic horses that partially replicate the work of wild horses, gomphotheres, glyptodonts, and giant sloths.[6] Many of the jicaro trees in Costa Rica today may be the consequence of the activities of horses over the last few hundred years. The same might be the case for old guanacaste trees (*Enterolobium cyclocarpum*, the national tree of Costa Rica), bay cedars (*Guazuma ulmifolia*), and monkey pod trees (*Pithecellobium saman*), all of which are megafauna fruit trees with fruits now sometimes eaten by horses or cows. Horses and cows are no substitutes for giant sloths and gomphotheres, but they can nonetheless fulfill a few of their missing roles, especially in areas where forests have been cut down and must be, seed by seed, restored.

As a result of these experiments, Janzen had both helped to re-grow a little more forest and convinced himself that he was right about the megafauna and their fruit. He moved on to other ideas. There were to be many. Meanwhile, he had made many new predictions that could be tested. Such tests, however, would not be his focus. As he said in an email, he "was not interested" in spending the rest of his life "slicing the salami ever more thinly." By this, we suppose that he meant that he had resolved the big-picture story of fruit and megafauna to his own satisfaction. Yet, there was still one big mystery. Janzen started his work on tree species with megafauna fruits by noticing trees of which the seeds inside fruits were dying beneath their mother trees. But why weren't such trees already extinct? Ten to twelve thousand years had passed since the extinction of the megafauna, and the individual trees he was examining were not ten to twelve thousand years old. They'd gone through many generations. Somehow enough of their seeds had survived, germinated, and grown into trees to allow them to persist. But how?

Perhaps some of the trees had grown up from seeds dispersed by cows or horses. But this didn't explain all of the species, some of which are not eaten by cows or horses. In addition, there was a roughly ten-thousand-year gap between the extinction of the megafauna and the arrival of the horses and cows from Europe. Others of the trees might rely on smaller animals, such as rodents or parrots, to carry their seeds.[123] Recently, one species of palm tree in Brazil has been shown to have evolved smaller fruits in populations where bird species with big mouths have gone locally extinct.[124] Some big fruits might also float to new sites, if they happen to fall into rivers and float well.[7] But could these modes of dispersal really account for all of the tree species with megafauna fruits? There was one more possibility.

In the wake of the extinction of the megafauna, human populations became ever denser in the Americas and elsewhere. As they did, humans may have begun eating ever more of the fruits that the megafauna once ate. Many of the fruits, such as the stinking toe and jicaro fruits, had evolved defenses against being eaten by smaller mammals, including primates. But these defenses were no match for stone tools. What was more, because the fruits had evolved to attract animals with big nutritional needs, they tended to be sweet, but also sometimes fatty or rich in protein. If ever there were forbidden fruits, it was these. They were fruits with strong aromas and, often, delicious pulp that for millions of years had been inaccessible to primates. Now, in the absence of the megafauna, they hung unpicked. And they could be opened with a sharp stone.

Maybe humans had taken on part of the ecological role once played by giant sloths, mastodons, mammoths, and the like. This hypothesis offers a simple, testable prediction. If humans played an important part in keeping these species alive, one would expect that the subset of tree species with fruits eaten by humans should tend to be more common today than are species with megafauna fruits that are not eaten by humans.

The commonness or rarity of different megafauna fruit tree species would have been difficult for Janzen to test in the 1980s because data on the distribution of tropical plant species were both poor in quality and poorly consolidated (and anyway, he didn't want to slice the salami). Such data are still relatively poor in quality (and miss any species that went extinct before the data began to be gathered, which is to say any time between ten thousand years ago and about 1920). Yet, they are now, at least, consolidated. Recently, several teams of researchers have used versions of these newly consolidated historical data to compare the rarity of megafauna fruits and other fruits. In doing so, they have shown that trees with megafauna fruits tend to be at greater risk of endangerment and extinction than trees with other means of dispersal.[125] For example, the Kentucky coffeetree (*Gymnocladus dioicus*), a species with megafauna fruit (giant pods), is now rare everywhere it is found, lingering here and there along rivers where its seeds happen to wash up.[126] But one group of researchers also discovered something else.

A team led by Maarten van Zonneveld, at Biodiversity International, a global research organization with a center in Costa Rica, gathered a list of tropical tree species native to the Americas known to have megafauna fruits. They then sought to divide those species into three groups: species not eaten by humans (based on studies of people living in forests throughout the Americas), species eaten by humans, and species eaten by humans and cultivated by humans. They then could calculate the geographic ranges of the average species in each of those three groups. If ancient humans helped to save megafauna fruit trees by eating them, spreading their seeds, and even intentionally planting them, van Zonneveld expected to find that the species people found to be delicious were more likely to have bigger geographic ranges than those that are neither delicious nor eaten.

Something like this appears to be the case for honey locust (*Gleditsia triacanthos*). Honey locusts are leguminous trees that, like the stinking toe tree studied by Janzen, have long, hard, bean-like pods. Inside those pods are seeds surrounded by a sweet pulp. Like stinking toe pods, the pods of honey locusts tend to rot where they fall. But the honey locust is nonetheless common, here and there, in patches, especially in western North Carolina and eastern Tennessee. Those patches tend to occur in habitats that are wet, habitats alongside rivers, even though the tree itself germinates best in slightly drier upland sites. Recently, Robert Warren, an ecologist at the State

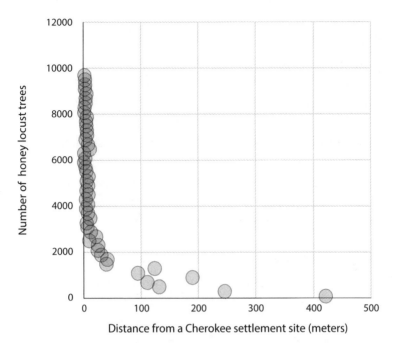

FIGURE 5.2. The density of honey locust trees as a function of distance from former Cherokee settlements.

University of New York, Buffalo State College, showed that these areas where honey locust is common are nearly all sites that were once Native American settlements, specifically settlements occupied by the Cherokee between roughly 1450 and 1840.[127] The Cherokee used the honey locust in drinks. They used the honey locust pulp like a sweetener. They found pleasure in the honey locust. In doing so, the Cherokee and their ancestors appear to have favored the survival of the honey locust. They may even have planted their seeds. As a result, where the Cherokee lived, the tree persisted. What if something similar happened with the most delicious (to humans) subset of megafauna fruit more generally? This is what van Zonneveld wanted to test.

No data on the taste preferences of ancient humans of the Americas exist, but van Zonneveld was able to find a database of the fruits eaten by communities living in and around the tropical forests of the Americas over the last hundred or so years. For example, the database recorded that the fruits of the stinking toe tree (*Cassia grandis*) and two of its relatives (*Cassia leiandra* and *Cassia occidentalis*) were eaten by humans. But it didn't say much more (for example, who ate the fruits, where, and for how long). The list would have to do. When van Zonneveld and his team compared the fruits that were eaten by humans to those that were not, they found that those that were eaten by humans had geographic ranges about one and a half times as big as tree species with megafauna fruits not eaten by humans. The tree species with megafauna fruits that are now cultivated by humans have larger geographic ranges still.[128] Conversely, tree species with fruits not eaten by humans, past or present, including some relatives of the stinking toe tree, have small, shrinking geographic ranges. Humans appear to have helped save many species of megafauna fruit trees by devouring them. Deliciousness saved these species. Or rather, our perception of deliciousness saved those species. Meanwhile, other species have suffered, their fruits producing aromas to lure giant mammals, day after day and year after year, giant mammals that never come.[8]

CHAPTER 6

On the Origin of Spices

The number of flavors is infinite, since every soluble body has a special flavor which does not wholly resemble any other . . .

—JEAN ANTHELME BRILLAT-SAVARIN

The pig flees from the oil of marjoram and fears every kind of unguent; for that which sometimes seems to give us new life is rank poison to the bristly pig.

—LUCRETIUS, *DE RERUM NATURA*

For the first three hundred million years of mammal evolution, our ancestors ate from among the species available to them. They chose some flavors over others, mammoths over howler monkeys, for instance. Some of our ancestors were choosier than others (they differed from each other as we differ from each other), but none could choose flavors that did not occur in nature. Nor do any among them appear to have mixed flavors, except inadvertently in the warm bowls of their mouths. Cooking offered new flavors, but the possibilities were finite. A mammoth foot roasted over an open fire could be roasted so that the meat was more or less moist and the skin more or less crackly, and yet it was inescapably, at the end of the day, always a mammoth's foot. A breakthrough in the story of flavor occurred when our ancestors began to add spices to cooked foods. In doing so, they took advantage both of the diverse chemicals found in plants and the human ability to learn to enjoy almost any aroma. They created new mixes of aromas and tastes. Then they learned to like those mixes.

As far as is known, no species except humans has learned to make a food by mixing multiple ingredients together. Chimpanzees don't add peas to their meat, nor do they add spices.[1] In addition, spicing food is not universal among humans. Some groups of people do not use any spices. Allen Holmberg, for instance, author of "Nomads of the Longbow," observed that the Sirionó of Bolivia used no spices of any kind in cooking.[129] The traditional diets of other Amazonian hunter-gatherer societies, including the Yanomamö, seem similarly spiceless. One of the few exceptions appears to be the use of the ashes of some plant species as a form of salt.[130] Amazonians are not unique in this regard. Many groups of people appear to have traditionally used no or only few spices in their foods.

We use the term "spices" here broadly to include the parts of plants used in food not for their nutritional value, but instead for some other reason, typically in small doses, and in ways that feature their aromas and flavors. Some spices, those that are often called "herbs," are the leaves of plants. Peppermint, spearmint, oregano, basil, bay, and lemongrass are all herbs. On many of these leaves one finds small, plump spheres in which the plant stores chemicals; when the spheres burst like small bombs, as in our mouths or when we cut or tear the leaves, the chemicals rise explosively into the air. Many spices, such as mustard, cumin, or anise, are seeds. Others, including chili peppers, black peppers, lemons, and limes, are whole fruits. Meanwhile, garlic, onions, and their many relatives are bulbs, cloves are flower buds, and saffron is the female sex part of a crocus.

The use of spices is a deceptively simple act. You add a little something else to your pot, pan, or bowl and, in doing so, alter the flavor of the resulting food. But it isn't really so simple. In most cases, you need a pot, pan, bowl, or at least some container in which to mix ingredients (though spices can also be, for example, rubbed on the surfaces of cooking meats). This can be remedied by digging a hole, sealing it, and using it like a vessel (into which a hot rock can be added to cause liquids to boil). But here is the other problem: essentially every spice that we employ today to flavor our foods is from a strong-smelling plant part, whether a bulb, a leaf, or a seed. Most of those plants evolved the ability to produce the chemicals that we perceive as strong smells to ward off their enemies.

Hundreds of millions of years ago the first plants colonized land. Much later, when the first animals crept ashore, those plants were relatively undefended. To the first landlubbing herbivores the Earth was an endless salad. But only for a moment. Plant species able to evolve toxic leaves and reproductive parts, such as seeds, were far more likely to survive. And so, they did.

Eventually, most plant species evolved defenses. Some were physical. For example, bits of silica in grassland plants deter even the biggest of mammalian herbivores. The silica gives those grasses a terrible mouthfeel; eating plants that contain lots of silica is like eating a bed of lettuce on which someone has sprinkled sand.[131] Many, though, were chemical. Plants evolved chemical defenses they used to punish herbivores with convulsions, vomiting, and death. Such defenses often played a dual role of both deterring herbivores and killing the plant's

FIGURE 6.1. The leaf of a mint. The large deflated spheres on the surface of the leaf are the small containers in which the plant, like most herbs, keeps its chemical weapons. The contents of these containers are released when the leaf is bitten, torn, or crushed.

pathogens. In response to these chemicals, individual animal species evolved countermeasures (as did some pathogens), including specialized abilities to break down some chemical defenses. Plants, in response, evolved new defenses. This back and forth war helped to engender much of the diversity of plants and plant-eating animals on Earth.[132] And the back and forth is ongoing. Many species and varieties of thyme (*Thymus*) grow wild throughout the Mediterranean. Different varieties of thyme produce different defensive aromas. Often, the variety on one hill produces a different aroma from that on a nearby hill. These differences, from one hill to the next, have been shown to be partially determined by which herbivore, or other enemy, is most abundant.[2] Where sheep are rare but slugs common, the thyme variety that produces an aroma that deters slugs thrives. Where sheep are common, the variety of thyme that produces aromas that sheep dislike is most common. Similarly, thyme basil,[3] which also grows in the Mediterranean, produces much less of its most aromatic compounds when growing in areas that are inaccessible to goats and sheep. It offers less warning when there is no one to warn. Based on these and other observations, some scientists have gone so far as to suggest that the aromas of plants in the Mediterranean and Middle East are an effect of the thousands of years of herbivory by goats and sheep. The species and varieties that are left are those that were most strongly defended.[133] The most geographically widespread thyme in Europe today, for example, is the variety that has the strongest chemical deterrents.[134]

The war between herbivores and plants never completely ended anywhere. It never will. Yet, our bodies have signed truces with many individual plant species and lineages. The truces are manifest in our mouths as bitter tastes. Animals, including our ancestors, evolved bitter taste receptors to warn them away from plants they were unable to detoxify. These receptors are a little different for each animal species, as a function of what they are and are not able to detoxify. Bitter taste receptors allowed animals an easy system for knowing what not to eat. We have a bitter taste receptor, for instance, that tells us to avoid plants with strychnine, another that tells us to avoid caffeine. Fifteen different compounds in hops trigger at least one of three human bitter taste receptors.[135] In return, the plants offered something. They evolved to produce aromas that are themselves not toxic, but that instead warn of the presence of toxins. Much as a monarch's coloration tells birds, "Don't eat me," so too do the aromas of some plants. Animals could then avoid toxic-smelling plants and, in doing so, avoid tasting bitter plants, increase their own odds of surviving, and leave the plant in relative peace.

To employ a spice is to ignore nature's admonishments. We humans intentionally gather plants with high concentrations of defensive chemicals or warning aromas and add them to our food, typically in small doses. The chemicals associated with the bitter tastes of dandelions and dill, for example, are poisons. The fragrant aromas of garlic, mint, thyme, and dill are warnings of the existence of poisons. They say, without any real ambiguity, "Go away you beast with terrible gnashing teeth and bad breath or I will make you suffer." Eating such plants despite their warnings is a bold act. Yet, it is one to which we have grown numb. We are so accustomed to the flavors and aromas of spices that we don't consider the unusualness of consuming them. With spices, then, we need to explain two things. We need to explain how it is that we humans so readily convince ourselves that spices are pleasing. Then we need to make sense of why we began to do so, why we began to spice foods and to enjoy spiced foods.

The first question, that of how we learn to enjoy spices, is actually the easier of the two to answer. Babies begin to learn to enjoy the aromas (and ultimately flavors) of particular spices in utero, and then this learning is reinforced once they are born.

During pregnancy, a fetus experiences the tastes and aromas of the foods its mother consumes. The chemicals from the food enter the amniotic fluid and travel into the nose of the fetus. The fetus can sniff at the tiny sea within which it floats. Fetuses appear to be predisposed to learn that the maternal aromas in which they swim are pleasurable, worth seeking out once born into the world. This is true even if the aromas are defensive compounds of plants. For example, when mother sheep eat garlic, their amniotic fluid smells of the defensive chemicals in garlic.[4] The fetal sheep smell the aroma. Having been exposed to it, they prefer it once they are born. If the amniotic fluid of pregnant rats is injected with garlic, the babies of those rats, upon birth, involuntarily try to suckle when presented with garlic; they pucker pink lips and searched for their mothers. "Where are you, my garlicky mom?"

In studies of humans, experiments have been less invasive, but the results have been similar. In one study, Benoist Schaal and colleagues at the French National Center for Scientific Research (CNRS) compared two groups of women in the Alsace region of France. One group was given as many anise-flavored mints, anise-flavored cookies, and anise-flavored syrup samples as they wanted during the last ten days of pregnancy. The other group was given no anise-flavored food and was asked to refrain from eating any anise-flavored food (and they appear to have complied). The researchers then tested whether the babies born to the two groups of women differed in their fondness for the compound that gives anise its aroma, anethole. Babies whose mothers did not eat anise tended to show faces indicative of displeasure at birth when offered a highly diluted sample of the anise smell, anethole.[136][5] The babies whose mothers ate anise, on the other hand, were more likely to turn their heads toward the anethole smell, stick their tongues out, and move their tongues as if to lick their lips.

In another study of humans, babies whose mothers had eaten garlic during pregnancy puckered their lips to suckle at the aroma of garlic. Similar effects of the womb's flavors have more recently been shown for peas, green beans, and sulfurous cheeses such as Camembert, Munster, or Époisses. Eight-month-old babies whose mothers ate peas, green beans, and other green vegetables when pregnant showed a preference for the aroma associated with the smell of greenness (2-isobutyl-3-methoxypyrazine). The eight-month-olds whose mothers ate sulfurous cheese when pregnant showed a preference for the smell of dimethyl sulfide (which is present both in sulfurous cheeses and garlic). Mothers who ate fish while breast-feeding tended to have babies that enjoyed the aroma of fish, or at least the compound associated with the aroma of fish, trimethyl-amine.[137] Trimethylamine is found in both the amniotic fluid and breast milk of fish-eating mothers. These effects of aromas associated with amniotic fluids and breast-feeding appear to sometimes, though not always, persist into childhood and later life.[138]

Nature tells animals, such as humans, to trust their mothers and the aromas associated with what they have eaten. In the small communities of our ancestors, the aromas associated with mothers' foods tended to be, with a few exceptions, the aromas associated with the food of the broader community.[6]

The net result of the olfactory learning we, as mammals, do before and just after birth is that we are able to accumulate knowledge about beneficial foods and dangerous foods across generations even without any teaching. Returning briefly to chimpanzee culinary traditions, prenatal learning might be sufficient to help teach baby chimpanzees many of the species that should be eaten, particularly those species with strong aromas. This would also have been true of the common ancestor of humans and chimpanzees six million years ago. It is true for us today, with an additional feature. Our ability to speak helps us to layer complexity upon a more ancient architecture of preference. Our mothers' bodies teach us what to love, and our parents' words remind us what to love. Both of those influences are added to by the actions and culinary offerings of the rest of our community, which remind us what our people love. As a result, it would have been relatively easy for our ancestors to learn to love spices and, at the same time, forget that it had ever been otherwise.

But when and why did some humans start to use spices in the first place? And what reason did they have to learn to love them?

Here and there in the archaeological record one finds evidence of what might be (and also might not be) spice use. Hackberries (*Celtis* sp.), for example, have been found in a 60,000-year-old Neanderthal hearth at Dederiyeh cave in Syria.[139] The hackberries of the region, like those in North America, are somewhat unpleasant and not terribly satisfying to eat on their own. Native Americans in the desert Southwest of the United States use similar hackberries as a spice. They add them to meat dishes while the dishes are cooking, as one might use peppercorns. Could Neanderthals could have been covering their meat with hackberries before cooking it, to add flavor? No one yet knows.

One of the oldest well-documented instances of the use of spices is surprisingly recent, from an archaeological site dated to no older than 6600 years ago. The evidence comes from a study conducted by archaeologist Hayley Saul, York University professor Oliver Craig (at the time Saul's advisor), and their colleagues in Spain and Denmark. The study considered a number of archaeological sites. But the most detailed work was done on a site in northern Germany from a time in which agricultural cultures were spreading north and hunter-gatherers were undergoing transitions in their food ways. The site, Neustadt, was first occupied by hunter-gatherers about 4600 BCE, and then continued to be occupied for another eight hundred years as those hunter-gatherers transitioned to agriculture. At this site, Saul, Craig, and colleagues could study the transition between hunter-gatherer lifestyles and agricultural lifestyles based on how both the ceramics made and the food eaten at the site changed through time. The earliest inhabitants of the site were hunter-gatherer people known for making large, ceramic vessels of a style called "Ertebølle" and so are referred to as the Ertebølle people. The later agricultural inhabitants made smaller ceramics of a type called "funnel beaker" and so are referred to as the Funnelbeaker people. (Based on this archaeological nomenclature, we would probably be the "plastic cup people.")

Saul, Craig, and colleagues were able to find Ertebølle ceramic vessels from the site that contained what archaeologists call "foodcrusts." These foodcrusts were evidence, first and foremost, that ancient northern Europeans were not very good at doing dishes. But they could also be used to study what these people were eating. The Ertebølle hunter-gatherer foodcrusts contained both meat and plants (whereas the later funnel beaker ceramics tended to be more specialized, containing either meat or plants). Saul discerned, using a variety of laboratory approaches, that the meat in the Ertebølle foodcrusts was from wild animals, roughly half marine and half terrestrial, say, from fish and deer. Meanwhile, Saul found that while some of the plant material was starch of some staple (perhaps, Craig speculated in an email, hazelnuts and acorns), much of it was from the seeds of garlic mustard (*Alliaria petiolata*). Garlic mustard is not related to garlic or leeks, but is instead a garlicky member of the mustard clan. Saul and Craig and their colleagues suggest that the garlic mustard they found caked in pots was being used as a spice. In other words, they appear to have found evidence of an ancient stew containing meat, fat, some starch, and a garlicky spice cooked by Ertebølle hunter-gatherers. The recipe seems to have been simple, something like this:

Take the meat of fish or mammal. Add it, along with bones and sinew, to water in a ceramic vessel. While it is cooking, add hazelnuts or roots. Mix in garlic mustard to taste. Share.

Inasmuch as Saul, Craig, and colleagues also found beeswax in some cooking vessels, it is also possible that honey was added to the recipe. In an email, Craig speculated that perhaps the Ertebølle hunter-gatherers began using ceramics so that they could make these sorts of dishes. "What if the main reasons for inventing/adopting the cooking pot was to allow flavors and textures to be combined in new ways, as part of a new culinary aesthetic?"

We suspect that the culinary aesthetic that included spices, whether in cooking pots or other contexts, likely began for different reasons in different cultures and with different spices. Some spice uses may represent idiosyncratic, aesthetic manifestations of culture. Others may have begun as a kind of culinary preventative medicine. Retired Cornell University professor Paul Sherman has argued that spices might have first been used in order to kill off pathogens present in food. They might also have warded off the establishment of pathogens in food left overnight or over a few days (or food inadvertently left on imperfectly cleaned vessels). Humans may have especially used as spices those plant parts that were good at preserving leftovers and had strong aromas.[140] This use of plants as spices might have been an extension of the more ancient use of plants as medicine. Even today, many spices are used as both medicines and spices. For example, a plant called bitter leaf (*Vernonia amygdalina*) is used by some people both as a medicine and as a spice in cooked foods. It is added, for example, to the Nigerian meat stew egusi.[141]7

Sherman's hypothesis matches up well with what is known about how we learn to like and dislike aromas, both in utero and after birth. As we've just noted, prenatal learning teaches humans to love many different kinds of aromas (and their associated flavors). But it is then complemented by the learning that occurs later in life. As mentioned in chapter 3, when we learn aromas we rank them in part as a function of whether the memories we associate with them are good or bad. An aroma with many good memories comes to be ranked as good. We might imagine then that, for example, if a fetus was exposed to anise in utero it might be born with a fondness for the aroma and flavor of anise. And, if during childhood and later life that same individual then had many positive experiences associated with anise, its newborn fondness would be reinforced. Conversely, aromas that we associate with being sick are learned, very quickly, to be negative aromas. Humans can learn to dislike an aroma in a single event if that aroma, for example, is associated with vomiting (a phenomenon known as the Garcia effect). Spices that kept food safe might then be learned to be pleasing in utero and later in life, whereas the aromas of the same dishes without spices, if they caused food-borne illness even once, might be learned to be bad.

Sherman's idea also leads to a series of predictions. Some are nuanced. For example, spices should be most likely to be used in the form in which their antimicrobial chemicals are most active. Spices whose chemical activity survives cooking might be used in cooking. Those whose activity is greatest when raw might be more likely to be used raw. But the simplest prediction is that spices should be able to kill the bacteria species in foods able to make us sick. At least in trials in petri dishes in which bacteria are grown in the presence and, separately, absence of particular spices, many spices can do just that. As can be seen in figure 6.2, laboratory studies show that *some* plants that are used as spices are antimicrobial. Or, at least at they can be antimicrobial at the concentrations used in laboratory studies

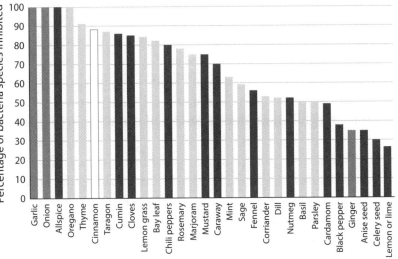

FIGURE 6.2. Percentage of food-borne pathogen species tested that each spice or its chemical compounds was able to inhibit. Spices are shaded as a function of the plant part from which spices are made. Grey bars are bulbs, roots or tubers. Black bars are seeds or fruits. The lone white bar is for bark. Light gray bars are leaves used as spices.

(conversely, others definitely are not, even at those concentrations). Among those spices with antimicrobial properties, garlics and other alliums (such as onions and leeks) are relatively well-studied examples.

Garlic and other alliums have a unique chemical arsenal that they rely on for defense. In garlic, this arsenal depends on two key compounds, alliin and alliinase. These compounds are stored in separate chambers in a garlic bulb and come into contact only once the bulb is damaged. Alliinase is an enzyme. When the bulb is bitten (whether by a insect, a rodent, or a human) the alliinase makes contact with the alliin and converts it in an instant into allicin. Allicin gives garlic its pungent aroma. Onions are similar except that a second reaction causes the allicin-like compound in onions to be transformed once more and become one of a variety of chemicals known as "lacrimators" (tear makers). Onions aren't the only allium to produce lacrimators; garlics do too. But onions produce more of the compounds than do other alliums. Once in our eyes (or those of a forest rodent), these lacrimators irritate nerve endings and further break down into sulfuric acid and other even more bothersome compounds.[8]

Despite their strong defenses against being eaten, garlics and other alliums are featured in recipes in many parts of the world. These plants are good candidates for spices that we have learned (from other mothers) to love, perhaps because of their antimicrobial ability. But no one had ever done the experiment of making an ancient dish with and without its garlic to see what happened. We decided we would. Recently, we worked with students at two high schools in Raleigh, North Carolina, to cook a dish called puhadi, with and without alliums.[9] We then studied what grew in the stew after it was left out at room temperature for several days. The recipe we used for puhadi came from one of a series of recipes written in cuneiform on a 3600-year-old tablet held in the Yale Babylonian Collection.[10][142] Recently, scholars at Harvard and Yale have reconstructed these recipes in a book titled *Ancient Mesopotamia Speaks*.[143] Nearly every one of the recipes features more than one allium. We chose puhadi because it featured four different alliums: onions, shallots, garlic, and leeks. The recipe reads:

> Stew of lamb. Meat is used. You prepare water. You add fat. You add fine-grained salt, dried barley cakes, onion, Persian shallot and milk. You crush and add leek and garlic.

Similar garlicky stews are also likely to have been eaten long before the time of ancient Babylon, both in the region of Babylon and beyond.[11]

The students made the puhadi recipe. They created replicate batches with and without alliums. They then watched what happened. To their delight, the version without alliums went bad (spoiled) very quickly and smelled terrible; that with alliums stayed more or less unchanged for several days.

The ways in which people have historically used alliums seem in line with what we might expect if some spices were first used as antimicrobials. Alliums are antimicrobial, even in the context of actual recipes. And humans can learn to love them even if our innate tendency (and the plants' intent for us) is to avoid them. But if spices were more generally associated with preserving food, we should expect to see some broader patterns too in where and how they are used. We would expect spice use to be common where conditions are hot and wet and pathogens grow quickly. This prediction is easy enough to make, but harder to test well. Sherman and a student, Jennifer Billing, tried one approach. They compiled recipes from around the world and then compared the mean number of spices in different recipes. What they found is that the hotter a region is, the more kinds of spices are found in the average recipe, as they had predicted (figure 6.3). However, there are other reasons this pattern might be expected. For example, it is possible that more kinds of plants with the potential to be used as spices can grow in warm, wet places. In theory, it should be possible to statistically distinguish these two explanations, but no studies have done so, as of yet.

Sherman and Billing also predicted that spices would be more likely to be used in meat dishes (in which food-borne pathogens are more common) than in vegetable dishes (in which pathogens would have historically been far less common). At least in their preliminary analyses, this seems to be the case; however, it is possible that the cultures and dishes represented in cookbooks are a non-random subset of all of those that exist in the world. Some of the peoples that are (or were)

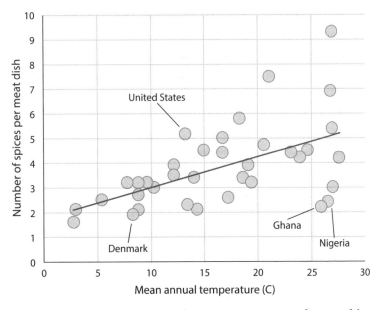

FIGURE 6.3. The number of spices used, on average, per recipe as a function of the mean annual temperature of the country in which they are used. Each point represents a country. Warmer places tend to have more recipes, particularly meat recipes, that include more kinds of spices. Countries below the line, such as Denmark, Ghana, and Nigeria, employ fewer spices than might be expected given their temperature. Countries above the line, including the United States, employ more.

most reliant on meat, such as hunter-gatherers of the Amazon or of the Arctic, use few spices or even no spices. But such cultures are underrepresented in cookbooks. Nonetheless, Sherman and Billing's analysis shows, at the very least, that it is common to use spices on meat.

Spicing of meat is not a novel feature of modernity. Consider the meat dish mentioned by the Roman Marcus Gavius Apicius (roughly 80 BCE to 20 CE) in his book *De re coquinaria*, "On Culinary Matters." Apicius wrote that to make the dish one need only

Pepper, lovage, parsley, dry mint, fennel, blossoms moist-
ened with wine; add roasted nuts from Pontus [a region in
what is now Turkey] or almonds, a little honey, wine, vinegar, and broth to taste. Put oil in a pot, and heat and stir the
sauce, adding green celery seed, cat mint; carve the fowl and
cover with sauce.

By our count that is at least seven spices in a single dish, a dish
that would have been flavorful even without any of the spices,
a dish so full of wild chemistry as to probably be safe from
pathogens for a long time.

There seems to be reasonable evidence for some relationship
between the use of some kinds of spices and the value of those
spices in controlling food-borne pathogens. People learned to
appreciate the value of some spices that they associated with
wellness. Conversely, they quickly learned to avoid spices and
other ingredients that tended to make them sick. The ability of
humans to learn aromas and assign them positive and negative
feelings was key to such learning.

Yet, this isn't the whole story. For one thing, in many cases,
spices that were once used as a kind of food medicine are now
used in regions in which such a use has very little practical
value. Imagine, for example, that you are in southern Italy enjoying a marinara pizza, shining with local olive oil and redolent
of garlic. That pizza has been cooked and is about to be eaten
promptly after having been cooked. The risk of food-borne illness from the pizza (which contains no meat) is remarkably
close to zero and probably in no way influenced by the garlic. It
may be the case that even when the use of a particular spice
begins because of its functions, it can readily come to take on
another role, that of adding a new dimension to food, something interesting and exciting. But we (and here we mean the

Western, societal we) love garlic so much that it is hard to imagine it as being anything but pleasing. And so perhaps it is easier to consider this phenomenon in the context of another spice, hops.

Hops are the flowers of the whimsically named plant *Humulus lupulus*. These flowers (or cones) were first added to beer as a food safety measure. We know this because it happened during the Middle Ages, and the reason for the use was recorded. The addition of hops helps to kill bacteria in beer that otherwise spoil the beer. As a result, traditional beers made with hops lasted longer and were better for shipping. But the flavor of hoppy beer was not originally appealing to consumers. Even today, it is not appealing to consumers who are used to drinking less hoppy kinds of beer. Over time, however, the preference for hops increased. Hops added a dimension to beer that was new and different. Some beer drinkers learned to associate hops with pleasure (this is one type of learning that we hope did not occur in utero, but it could have). Now, the value of hops in preserving beer is modest (beer making has many approaches to keeping bad-news bacteria at bay), and the reason for its inclusion is the unique flavor of the beer, a flavor that is even a little bit bitter, a flavor that is a warning, a flavor that we like despite the fact that it is telling us to go away.

Stepping back, we might then imagine a scenario in which the functional value of some spices is more important when conditions are hot and wet (and pathogens grow quickly), where refrigerating food is impossible, and in a number of other circumstances in which pathogens are particularly likely to be a problem. But we might also expect spices to be used more often in regions and cultures where foods are bland and where adding a new dimension to the culinary experience is valued. This would have been the case as crops were being domesticated and human diets, particularly urban diets, were becoming less diverse and more dominated by individual grains such as rice, wheat, millet, or corn. In other words, spices may have been yet another tool for making the food at hand more pleasurable than it might otherwise have been. In this telling, the use of spices to flavor a simple rice dish is not so very different from the use of tools, by chimpanzees, to access and eat ants. While this seems very plausible, it is a possibility that is remarkably difficult to study in a historical context. Maybe the easiest way to consider the role of spices in the pleasure of food is by considering those spices that are not antimicrobial.

In figure 6.2 you will notice that while a number of spices are very antimicrobial, not all are. Among the spices that seem to have very little value in preserving food and keeping food-borne pathogens at bay is black pepper. Black pepper is an iconic spice of Europe, ancient and modern. It was black pepper, among a handful of other spices, that sent Columbus searching for a new route to India. At various points in history, black pepper was worth considerably more than gold. Yet, black pepper is now regarded by experts in food safety, such as my colleague Ben Chapman, as a potential *source* of food-borne pathogens.[144] Some pathogens hang out, happily, among the cracks and crevices of peppercorns. The use of black pepper does not appear to be about controlling food-borne pathogens. Its use may have begun as a way to add a new dimension to the flavor of foods. It has a distinct aroma and flavor. However, it also possesses another kind of effect.

Black pepper is one of several spices that trigger a kind of receptor on the tongue associated with a wholly different sort of taste, a taste so unique it gets the opaque and somewhat un-inspired scientific name, *chemesthesis*. Chemesthesis occurs when chemicals in food trigger receptors associated with pain or touch. The active ingredient in black peppercorns is piperine (*Piper* is the genus of black pepper; it is also the word, in Latin, for pepper). Piperine is a perfect match for a receptor in our mouths, TRPV1, that evolved to be responsive to actual temperature heat. If you put coffee that is too hot in your mouth, it is TRPV1 that sends a signal to your brain. Your mouth is burning! The piperine in black pepper lands on these same receptors and triggers their excitement exactly as if it were fire. Black pepper tastes hot because piperine's "key" fits TRPV1's lock. In doing so, it tricks your mouth into believing it is experiencing real heat. Your body responds exactly as if you had accidentally mouthed a hot stone.

The piperine in black pepper is not the only chemical that triggers TRPV1's lock. Capsaicin, the active ingredient in chili peppers, does as well. Nor is that the end of the story. A compound present in the bark of cinnamon has an effect very similar to that of capsaicin or piperine, albeit milder.[12] It binds to the same receptor as do capsaicin and piperine. Chemicals in horseradish, wasabi, and mustard all also bind to that receptor. They do so more in the nose, however (where humans also have TRPV1 receptors), than in the mouth (such that they make your nose tingle in addition to making your tongue burn). Or, if you eat a little too much, they make your mouth burn and your nose tingle and burn. Something similar, albeit sort of the opposite, happens when we eat mint. Most mints, including spearmint and cornmint, contain menthol. Menthol has an aroma that we smell, but once in our mouths menthol also binds to the receptor that senses cold (TRPM8, if you are keeping track). And so, menthol makes our mouths feel cool. Sichuan peppers, which are not related to black pepper or to chili peppers, trigger the heat receptor (TRPV1), but they also trigger KCNK and TRPA1 receptors, which, for some reason that is not yet understood, upon being triggered, produce a tingling sensation.

At least some of the plants that produce these compounds appear to do so in order to help steer their fruits to one disperser rather than another. Such is the case with chilies. Birds have a receptor in their mouths that detects heat, but it is a tiny bit different than that of mammals, different enough that capsaicin does not trigger its activity. As a result, when birds eat chilies, they do not experience any heat. It appears that chilies evolved to produce capsaicin in their fruits, in part, so as to steer their fruits toward birds. Chilies without capsaicin tend to get gnawed on by rodents, who aren't very likely to carry their seeds very far. But chilies with capsaicin are avoided by rodents. Rodents aren't conscious enough to know that the burning they are feeling in their mouths poses no real danger. Meanwhile, birds don't experience the heat at all. They pick up the fruits, ingest them, and then carry them away to other fields, where the seeds are "planted." As an added bonus, chilies with capsaicin are also better defended against fungi. Chilies with capsaicin are more likely to get where they are going and more likely to survive once they get there.[13]

None of this, however, explains why humans use chilies or black peppers as spices. It instead tells us something of the opposite, namely that plants produce the active ingredients in chilies, black pepper, and probably other spices that trigger similar effects not just as a warning signal, but specifically as a warning signal to our kind. "Mammal, go away." One explanation for why we use these spices is that they offer a specific new dimension to food, that of culinary danger.

We jump off bridges to test the bungee rope. Thanks to chemesthesis, we put many things in our mouths that seem like danger, but aren't, for a daily version of the same sort of thrill. This is the hypothesis suggested by evidence gathered by the psychologist Paul Rozin, evidence gathered by studying pigs, dogs, rats, humans, and two chimpanzees. Rozin focused on chilies, though he might have just as easily studied black peppercorns or Sichuan peppers.

In one study on humans, Rozin decided to try to understand how the spiciness of a pepper related to the perceived tastiness of the pepper. He selected a group of individuals that included both people who enjoyed spicy food and those who did not. He offered those people one cracker after another made with the capsaicin from chilies. As he did, he slowly increased the amount of capsaicin in the crackers that they were offered until people said, "No more." Then, he asked people which cracker was the tastiest. They might have simply disliked all of the spicy crackers. Or they might have all thought the same level of spiciness to be the tastiest (that level at which food preservation was most effective). Or their preferences might have been all over the board, random. None of these was the case. Instead, people tended to choose as tastiest the spiciest cracker they could tolerate. They liked a level of spiciness that was just shy of agony. This is the behavior that would be expected if people ate peppers to enjoy the danger's biochemical pleasures. Pain and fear tell us, respectivelly, to stop what we are doing and run. But they also trigger the release of endorphins and other brain chemicals. Perhaps eating chilies gives us the high of fleeing danger without the bother of exertion or the threat of actual death. Rozin's sample size for this study wasn't enormous, but his results were interesting. It was on the basis of this study, and others like it, that Rozin suggested that chili peppers are well loved because they seem dangerous but aren't; they offer us what he has called "benign masochism."[145] This benign masochism, he argues, is a unique feature of humans. In short, Rozin thinks that we are simple enough to enjoy the consequences of hurting ourselves a little and smart enough to know that the hurt is not real and will eventually go away.

To like chilies, Rozin argues, a mammal has to learn to ignore the signs of danger, to know the signs are false. Rozin hypothesizes that this ability might be either uniquely human, or, if not uniquely human, at least confined to humans and species that learn to trust humans.[14] Obviously, non-human animals of many sorts learn, but ordinary learning may not be sufficient to acquire a love of chilies. Learning to love chemesthetic spices may require extraordinary self-awareness or extraordinary trust. Rozin decided to explore these ideas further with pet dogs and pigs. He would test whether pet dogs and pigs could learn to enjoy spicy food, whether through self-awareness, trust, or some mix thereof. Dogs are well known for being able to learn to love many aromas. So too pigs. But both are, by any measure, less self-aware than are humans. If learning to love spices requires the knowledge that their fire is false, then pigs and dogs might not enjoy spices even when their daily food tends to be spicy.

Rozin traveled to a village in the Oaxaca region of Mexico where nearly all food is spicy and where nearly all leftovers given to dogs and pigs are also spicy. Rozin asked twenty-two people whether or not their domestic dogs and pigs preferred spicy food. Despite the ridiculousness of the question, people answered Rozin. Just two out of twenty-two owners interviewed said that their dogs or pigs preferred food with chili peppers; in both cases, they were referring to dogs. The two dogs that were reported to like spicy food were then experimentally given food with and without chilies. They were equally likely to choose one dish or the other. They didn't prefer chilies, they just didn't care. Twenty dogs disliked food with chilies, two just didn't care.[146] This is what you would expect if part of what is required to appreciate chilies is not just the sense that they are associated with good food, but also the awareness, the conscious awareness, that what seems like dangerous pain is just a kind of mirage of the mouth. As an additional test, Rozin tried his chili experiment with rats. He raised one group of rats on chili-spiced food from birth. He raised another group on chili-free food but then later slowly began to add chili to their food. Both groups of rats had ample chance to learn to like chilies, whether from birth or slowly thereafter. Yet, both groups of rats, when given a choice between chili-spiced food and food without chilies, still preferred the food without chilies. The data seemed to suggest that rats are unable to learn that chilies are good. To be sure, Rozin upped the ante. He gave the rats both food without chilies and food with chilies, but he laced the food lacking chilies with a compound that would make the rats vomit. Then he tested the rats' preferences. The rats still preferred the plain food even though every time they ate it they vomited. Rats, like dogs and pigs, seem unable to learn to like chilies.[147] In other news, be cautious when offered food by Paul Rozin.

In general, mammals seem unable to learn to like chilies, with two exceptions. One exception is humans. The other exception seems to be a small subset of captive mammals that are either clever enough to realize that the pain of peppers is not real or trusting enough of the person who offers the spicy food to know that it is likely safe. The list of such mammals so far includes two chimpanzees cared for by humans, two pet macaques, and one very trusting American dog named Moose.[148] Rozin has not repeated his experiments with black pepper, Sichuan, or mint, but their stories seem likely to be similar.

Returning to the broader story of spices and humans, we think that as we conduct more studies we will find that spices have served multiple roles in human history and prehistory, just as the compounds in spices play multiple roles in nature. Once humans began to store food for longer periods and settle more permanently—before the origin of agriculture, but probably not much before—some spices may have been added to food to help to keep that food safe. The subconscious learning that goes on in human noses and brains made it easy to learn to love spices that helped keep people safe. Some spices also added pleasure to food, which would have been a particular benefit where settlements were becoming larger and the best-tasting species were becoming scarcer. The pleasure was, in some cases, a pleasing taste or flavor, complexity, in others, a thrill. The health benefits, flavors, and thrills of spices would all increase as crops were domesticated and settlements became larger (and more prone to food-borne diseases) at the same time that the average dish was more likely to depend on a relatively bland staple such as rice, cassava, corn, or wheat. Once spices became common, they could then be subject to all of the vagaries of history. Some would become expensive by virtue of their rarity. Others would come to be viewed as magical, sexual, or some complex mix of the two. But all of these spices are based on the chemicals associated with the struggle of plants for persistence; their chemistry, however we use it, is a chemistry of defense, war, and reproduction, a chemistry that is just beginning to be understood and yet rises up from nearly every dish we consume.[15]

CHAPTER 7

Cheesy Horse and Sour Beer

Give strong drink unto him that is ready to perish, and wine unto them that be of heavy hearts. Let him drink, and forget his poverty, and remember his misery no more.

—PROVERBS, 31:6

While writing this book we had the chance to sit and talk to many scholars about many aspects of food. Often, upon talking for a while, it became clear that some big gap in humanity's collective understanding existed. Sometimes together we could begin to, if not close the gap, at least narrow it a little. Such has been the case with sour taste.

We did not describe sour taste in any real detail in the first chapter on taste because it is different than the other tastes. In humans, sour taste is not attractive in the simple way that is, for example, sweetness. Nor is it as aversive as is bitter taste. We can learn to enjoy some foods with bitter tastes, but we tend to only do so with age and only in association with foods that offer some other reward—bitter chocolate, bitter tea, bitter coffee, bitter hops. Sour is something a little different. Babies are born able to respond to sour taste (they pucker their lips).[149] Most children love sour taste; responses are variable among adult individuals and cultures. Some of this is learning. Some is genetic. Our feelings toward sour are nurture and nature in a mix that is hard to disentangle.[1] Nor is it understood why we have sour taste receptors in the first place. All of the existing explanations for these receptors are frustratingly incomplete.

One hypothesis is that sour taste evolved so as to prevent animals from ingesting acidic foods that might harm them. This could be true, but foods that are sufficiently acidic so as to cause harm to the animals that ingest them are relatively rare in nature. They include some fruits, stomach acid, and here and there the occasional volcanic hot spring. Little else. Also, this hypothesis assumes that sour taste is aversive, like bitter taste, but in at least some species, such as humans, it is not always aversive. A second hypothesis relates to vitamin C. Paul Breslin, an expert on the evolution of taste at the Monell Chemical Senses Center, has argued that sour taste leads some animals to seek out fruits that are sour and contain vitamin C. Vitamin C is also known as ascorbic acid. On its own, it can make substances sour. Some species of wood sorrel (*Oxalis* spp.), for instance, are sour to humans because of their mix of ascorbic acid and oxalic acid. Being able to detect vitamin C in foods is especially valuable for the animals, including primates, that cannot produce their own vitamin C, especially those living in grasslands where vitamin C can be hard to come by.[150] This hypothesis is intriguing but relates only to the subset of species for which sour tastes are attractive. A smattering of other explanations for the evolution of sour taste exist, but none of them have been studied in detail and none of them are more convincing than either of the two hypotheses just described. Sour taste is a mystery. It is a mystery we will not resolve here. However, we will provide some insight into the role sour taste may have played during the last two million years of our prehistory.

Last year Rob was invited to a weeklong meeting, sponsored by the Wenner-Gren Foundation, in a palace in Sintra, Portugal. At the conference scholars from a half dozen fields and as many countries spent a week talking about fermented foods, eating fermented foods, and drinking fermented drinks. It was, at least by Rob's account, wonderful. It was there that Rob met Katie Amato, a primatologist at Northwestern University. It was through Rob's conversations with Katie that we began to understand one aspect of the role sour may have played, if not in all of mammal evolution, at least during the last thirty million years of primate evolution. At the meeting, Katie presented an eye-opening talk about the origins of fermentation. In it, she hypothesized that hominins may have started to intentionally ferment fruits millions of years ago. It was in thinking about this hypothesis that some of the consequences of sour taste receptors (if not their origin story) began to become clearer.

To microbiologists, fermentation is any microbiological process that converts carbon compounds to energy, typically in the absence of oxygen. In the context of human food, we tend to consider fermentation to be the subset of those conversions that yield foods and drinks that humans ingest such as sour beer and sauerkraut, miso and sake. Fermentations of grains, roots, fruits, and other plant parts come in an extraordinary diversity of forms. But two common fermentations are those that yield acidic foods and drinks and those that yield alcoholic foods and drinks. The former tend to be dominated by lactic acid bacteria and acetic acid bacteria, the latter by yeasts. In practice, many fermentations, particularly those that rely on wild microbes, are a mix of the two and so both sour and alcoholic. Such is the case, for example, for sour beers, kombucha, and sourdough bread.

In one model of the history of food, a model first advocated by the botanist Jonathan Sauer, the first species domesticated by humans were the microbes used in making sour beer and sourdough bread. Then, once those microbes were domesticated, humans found themselves in need of a more reliable source of fodder for those microbes. Then, and only then, they began to domesticate grains in order to feed the microbes in order to make beer. In this telling, microbes are central and crops are secondary.[151]

Part of this perspective and chronology appears as though it may be right. At a 13,000-year-old hunter-gatherer archaeological site in Israel, Stanford archaeologist Li Liu and her colleagues found holes carved into the bedrock and, separately, into boulders. The holes in the boulders were used to store baskets filled with grains and other plant materials; the baskets were then covered with stones. Liu and colleagues believe that the holes in bedrock, meanwhile, were used to ferment a kind of barley beer. The beer would have likely been, Liu speculates, somewhat sour, and low in alcohol.[2] Once fermented to taste, it could have been be scooped out of the hole with a small container or even cupped hands.[152] This site in Israel is the oldest potential evidence of barley fermentation yet discovered and is from a time before the first evidence of agriculture.

The evidence for fermentation at the site in Israel is somewhat contentious among archaeologists (it is hard to unambiguously show that ancient stone vessels were used in brewing). Yet, even the archaeologists we talked to who were skeptical about the site thought it plausible that beer was being brewed thirteen thousand years ago. And scenarios in which brewing precedes agriculture are also plausible in other regions. University of British Columbia anthropologists John Smalley and Michael Blake argue that in the Americas corn stems appear to have been used to make a fermented, alcoholic drink before corn itself was domesticated. The stalks of the wild relative of corn, teosinte, appear to have been valuable for their fermentable sugar before they were valuable for their kernels.[153] And, as Smalley and Blake point out, it seems unlikely that corn stems were the first thing Native Americans tried to ferment; it is far more obvious to ferment a fruit than a stem.

Barley, corn, and rice might even have really been domesticated in part to provide larger quantities of sugars with which to make fermented drinks. The drinks offered our ancestors the advantage of clean, nutritious liquids, at least compared to the alternatives. And the alcohol in the drinks both pleased them and left them wanting more. Fermentation, in this story, is something that begins before agriculture but becomes elaborated with agriculture. In this telling, the invention of fermentation fits snugly between the first use of spices and the origin of agriculture on the big timeline of food and flavor. At her talk in Portugal, Katie Amato described why she thinks this telling is wrong, or if not wrong, incomplete.

Katie earned her degree chasing howler monkeys through Central American tropical rain forests (in order to catch their feces to study their gut microbes).[154] She excels both at paying attention to observations about the biology of primates that others miss and also putting those observations in the context of the broader primate story. It was one of those observations that she focused on in her talk.

In preparation for her talk in Sintra, Katie had asked researchers around the world about their experiences observing

primates interacting with fermented foods. One of the researchers who responded was Liz Mallot. Liz had been working in
Costa Rica when she observed something unusual involving
white-faced capuchin monkeys (*Cebus imitator*) and the giant
fruits of the almendro tree, *Dipteryx panamensis*. To understand
Liz's story, it is useful to know three things about almendro
trees. They are very tall, often a hundred feet or more. They
produce large fruits that appear to have evolved to be dispersed
by megafauna species, such as giant sloths, that would have
gathered fallen fruit from the ground.[155] And, finally, the trees
produce these fruits every other year, such that in any given year
they either have many, many fruits or none.

The day that Liz started to piece things together started like
other days. Liz ate a breakfast of gallo pinto (yesterday's beans
and rice mixed and fried) and scrambled eggs. Then she headed
out to find and watch the monkeys. When she found them, they
were near an almendro tree. The tree was full of fruits (it was
one of those years). But having evolved to be dispersed by
giant, extinct mammals, the fruits have a husk that capuchin
monkeys cannot usually break into with their teeth. The monkeys with the biggest jaws can sometimes bite into them, but
only sometimes, and, typically, with what looks like displeasure. And yet, Liz watched as some of the adult monkeys
climbed a hundred feet up to the top of the almendro tree and
started throwing the almendro fruits down. It is easy to notice
when monkeys are throwing giant hard-husked fruits down to
the ground, especially if you are standing beneath the monkeys.
Gravity is rarely on the side of primatologists.

It isn't uncommon for monkeys to knock fruits to the ground,
but to go to so much effort to do so was unusual, especially
given that the capuchin monkeys could not eat the fruits of
their labor. Eventually, the monkeys tired and climbed down

from the top of the almendro tree. By the time they did, the ground was covered with fruits, fruits the monkeys could not eat. The monkeys rolled a few around, whistled back and forth at each other, and moved on. But they did not move on for good. As Liz would realize in the coming days, they moved on for the time being. Days later, once the fruits had begun to rot, the monkeys came back to the almendro tree. When they did, they inspected the fallen fruits. If, and only if, the fruits were well rotten—the brown peel having darkened and worn away, exposing the fuzzy green pulp—the monkeys ate them.

Liz saw this chain of events, fruit knocking, waiting, and return to the rotten fruit, no fewer than three times. The only conclusion she could come to was that the capuchins were intentionally knocking the fruits down so that they would ferment on the ground. Once the fruits had fermented, they were softer and easier to digest. They were likely also a little sour due to the actions of lactic acid bacteria, and they had a nice aroma that Liz characterized as akin to fermented beans.[3] They may have been a tiny bit alcoholic due to the actions of yeasts.[156] They were, in essence, little bowls of food stuff akin to kombucha, seasoned with the cumin-like aroma of the seed inside the fruit (the seed is used by peoples throughout the region as a spice). In short, Liz thinks that the capuchin monkeys have figured out how to eat a fruit that evolved to be dispersed by megafauna by fermenting it. They use microbes as a tool.

On the basis of Liz's observations and those of other field primatologists, Katie Amato thinks that our ancestors have likely been fermenting fruits for millions of years. If Katie is right, the evidence of early fermentation in Israel, if it is verified, is simply evidence of when fermentation became sufficiently elaborated so as to require larger and longer-lasting vessels. One gap in this idea, a minor but intriguing one, is the question of

FIGURE 7.1. A white-faced capuchin monkey enjoying a fermented almendro fruit at Liz Mallot's field site in Costa Rica. Note the tonsured head of the capuchin, part of the reason for its common name in English.

how the capuchin monkeys are able to figure out that a fruit is safely fermented. Initiating the fermentation of fruits is no more difficult than other kinds of tool use, for example placing a palm nut on a flat, anvil-like rock and then smashing it open with another, well-chosen, rock (which another species of capuchin monkey has learned to do).[157] An animal smart enough to do one could also do the other. The challenge is microbiological. Fermentation is rot. The global food safety system is built around controlling rot. Foods that rot the wrong way can be very dangerous. Foods that rot the right way (from a human perspective) yield beer, bread, kombucha, and ham. How can a capuchin monkey, or how might a human ancestor many millions of years ago, tell a safely rotten food from a dangerously rotten one? Old jungle fruits do not have "best by" labels.

One possibility raised by Katie at her talk was that primate species judge the safety of fermented foods based on their acidity. Some fruits are acidic even when ripe. Think of lemons, plums, wild apples, or even grapes. But once fruits begin to rot, even sweet fruits can become acidic, if their fermentation is dominated by bacteria, or rather, certain bacteria. The acid in rotten foods, whether they are meats or fruits, is produced by one of two groups of bacteria, either lactic acid bacteria (which produce lactic acid) or acetic acid bacteria (which produce acetic acid). Both lactic acid and acetic acid bacteria produce acid as a defense in order to kill off their competition. The species with which these bacteria compete include the species that are dangerous to mammals. As a result, acidic foods are almost always free of pathogens. This reality, long understood by people who make sourdough bread, pickles, or sour beer, led Katie to posit that perhaps primates might use sour taste as a way to distinguish good rot from dangerous rot. Using sour taste as a judge of safe rot would be especially easy for species of primates for which sour taste is pleasurable. This isn't to say that sour taste, or even preferences for sour tastes, evolved in the first place for use in sampling the safety of fermented foods, but rather that it evolved for other reasons and then could have been co-opted as a kind of bodily pH strip in the service of fermentation.

The trick was that no one had compiled any sort of list of which species use their sour taste receptors to detect acidity, much less whether those species like or dislike that taste. Unfortunately, while the gene that controls the sour taste receptor (OTOP1) has recently been discovered, that gene plays multiple roles in the body. For example, it is involved in vestibular function, which is to say, balance. As a result, while changes in the gene could be studied, doing so would need to be interpreted with caution; they might not have anything at all to do with sour taste.

Upon return from Portugal, Rob decided to try to approach the question the old-fashioned way, by beginning with a thorough review of old studies from diverse fields of science. How hard could it be? At Rob's prompting, Hannah Frank, a student in a class Rob was teaching on flavor, began to compile literature records of animals that could taste sour. She then also identified which subset of those species liked sour tastes and which did not. The results were both clear and surprising. The clarity was with regard to which animals can taste sour foods. All of the mammals, birds, fish, and amphibians tested to date appear able to detect acidity using their sour taste receptors. It seems likely that vertebrates have been able to taste sour for hundreds of millions of years, since before the first fishy ancestor of all terrestrial vertebrates crawled ashore. Or at least this is what the story looks like so far. Many groups of mammals and birds have never been tested. For example, no studies appear to have considered any carnivores or carrion feeders, with the exception of a single very old study of domestic dogs, the results of which were ambiguous. Yet, Hannah found some thirty animal species for which the ability to detect sour taste had been studied, a pretty reasonable list. The surprise had to do with the response of those animals to sour tastes. Nearly all of them, twenty-six out of thirty, found even slightly acidic foods to be aversive. They turned down sour food even when there was nothing else; many turned down sour food even when it was both sour and sweet. This was true of mice, rats, cows, goats, sheep, black-handed tamarins, squirrel monkeys, and a couple of dozen other species. But there were exceptions.

One of the exceptions was the domestic pig. The ancestors of pigs were omnivores. They foraged for anything they could find and they foraged very often on the ground, where they might be more likely to encounter rotten fruits than fresh ones.

The second exception was pigtail macaques, which tend to have diets similar to those of wild pigs.[158] The third exception was night monkeys.[159] Night monkeys forage for fruit at night. They thus appear to rely on aroma to find the fruit and so might eat more rotten fruits than do other primates, if such fruits are easier to find in the dark.

The fourth exception was humans. Humans either innately like sour tastes or find it very easy to learn to like sour tastes. Even sour tastes associated with high concentrations of citric acid (oranges), acetic acid (vinegar), or lactic acid (sauerkraut) are pleasing to humans. All of this made us wonder about gorillas and chimpanzees. If gorillas and chimpanzees like sour foods, one might imagine that so too did our common ancestor. If they don't, then perhaps the human preference for sour foods is something recent, whether it is recently learned or a more recent evolutionary change. But Hannah's database did not have any data for chimpanzees or gorillas. Nor could we find anything in the literature. It was known that chimpanzees could detect sour tastes (based on studies of the responses of chimpanzee brains to acidic foods) but not whether they liked or disliked those tastes. We emailed a dozen chimpanzee researchers and no one knew of a test where chimpanzees were presented with foods varying in their degree of sourness to see what they might prefer. Then Rob sent a text to Christophe Boesch (actually, he sent a text to Mimi Arandjelovic, who sent it to Christophe, who sent his response back to Mimi, who sent it to him).

In response to Rob's question as to whether chimpanzees like sour food, Christophe wrote, "THEY LOVE IT!" He then went on to point us to the paper showing the fondness of chimpanzees for lemons. We then went back to Toshisada Nishida's study of the tastes of chimpanzee fruits and noticed that many of the fruits the chimpanzees preferred were either sweet and sour or sweet and very sour. In discussing sourness, Nishida mentioned the earlier studies by Jordi Sabater Pi in Equatorial Guinea, which found that both chimpanzees and gorillas like foods that are extremely sour. What was more, all of the main fruits eaten by chimpanzees in savanna habitats, such as those at a site called Fongoli in Senegal, are sour or sour-sweet (or at least they are sour-sweet to the human populations that also eat them).

It seems likely, in this context, that chimpanzees and gorillas are either hardwired to like sour foods or learn to do so very easily. Sabater Pi suggests that this fondness may relate to the increased time that chimpanzees and especially gorillas spend on the ground. It is possible, he argued some fifty years ago, that in spending more time on the ground, chimpanzees and gorillas have less access to fresh fruit. To them, fallen fruits are more apparent and easier to gather than those hanging on the tree, and fallen fruits are more likely to be rotten. For an animal species relying on fruits that are slightly rotten, being able to use sour taste to choose those that have been rotted by lactic acid bacteria or acetic acid bacteria might be advantageous.

Coming back to Katie's fermentation hypothesis, if ancient humans already had a fondness for sour tastes, it would have been especially easy for them to figure out how to safely ferment fruits and, in doing so, to make drinks or foods with aromas and flavors akin to those present in kombucha, aromas and flavors that they could learn to associate with health and pleasure. One might even imagine that individuals with a greater fondness for sour tastes might be more likely to survive. If that fondness were genetic, they might also be more likely to pass on the genes for that fondness. But as Katie pointed out in the talk she gave in Portugal, there is something more. In fruits and other rotting foods, lactic acid bacteria and acetic acid bacteria compete with other bacteria. They also compete with yeasts.[4]

Yeasts eat sugar. They do so by converting one molecule of glucose $(C_6H_{12}O_6)$ to two molecules of carbon dioxide (CO_2) and two molecules of ethanol (C_2H_5OH), aka alcohol of the sort found in cider, beer, and wine. This biochemical magic generates energy for the yeast. The ethanol is then excreted from the yeast cells as waste. Yes, your booze is fungal feces. But it needn't be. Yeasts can actually get more energy from nectar or fruits if they don't produce ethanol and instead break down the sugar more completely. So why produce ethanol? Just as bacteria produce acid to kill other bacteria and yeasts, yeasts produce alcohol to kill bacteria.[160] As a result, fruits and other substrates that are alcoholic also tend to be safe for animals to eat. The alcohol kills pathogens just as well as does acidity—but with a catch. Alcohol kills most bacteria (one interesting exception being acetic acid bacteria, which have evolved the ability to turn this toxin into energy and, in doing so, make vinegar), but it also makes most mammals, including most primates, sick.

For most primates, even a relatively small amount of alcohol consumption can lead to intoxication (a dangerous proposition when up in a tree). The products of alcohol metabolism, including acetaldehyde and acetic acid, build up in the liver. Although it is not yet well understood, this buildup may make wild primates feel unwell; they may suffer from nausea, headaches, and, well, an old-school simian hangover. If they do, this would seem to rule out the reliance of our ancestors on alcohol as an additional sign of safely rotted food. But there is slightly more to the story. Chimpanzees, gorillas, and humans have livers that are supercharged for breaking down alcohol into forms that are nontoxic. In the livers of all mammals, alcohol is converted to acetaldehyde by an enzyme called alcohol dehydrogenase (ADH).[5] That acetaldehyde is then converted into acetate by another enzyme. The chimpanzee, gorilla, and human versions of the enzymes that carry out these processes in the liver are much faster, some forty-fold in total, than those of other primates. As a result, these species can safely ingest more alcohol and, in so doing, rely on its calories and benefits with potentially fewer negative consequences. At some point, human ancestors also began to experience a feeling of euphoria when drinking alcohol. It is as of yet unclear when this response to alcohol evolved and whether it is an adaptation or simply an accident of the details of the complex interactions between alcohol and the brain.

Pulling this all together, monkeys can apparently figure out how to ferment fruits. So too, Katie Amato imagines, could ancient humans. In addition, ancient humans were likely able to enjoy not only the sour tastes of the fruits they had fermented and their complex aromas, but also their alcohol. They had the potential to be led by their tongues, noses, and brains to fruits made sour by *Lactobacillus* bacteria and made alcoholic by yeasts. Katie thinks that the forest forays of hominins into fermentation may have occurred during one of the periods in which they began to move out of the canopies of contracting forests and onto the ground of woodlands and savannas, where fresh fruits were rare and the ability to ferment less-edible fruits and roots may have been advantageous. This is roughly the same time period when our ancestors evolved their supercharged ability to metabolize ethanol.[161] The effective control of fermentation might have even been (Amato speculated one night while we sat around tasting port wines, their sweetness, richness, and slight sourness pleasing our mouths, while their chemistry pleased our minds) one of the critical inventions that led *Homo erectus* to be able to find enough energy to fuel their big brains and jettison their big jaws and teeth.

Whenever it occurred, the first intentional consumption of rotten fruits, roots, and stems would have improved the experience of the food, its tastes, flavors, and pleasures. This would have been true with regard to many fruits, but especially with regard to roots and stems. As with the case of the capuchin monkeys, the rotten foods would have been easier to chew and more pleasing to chew. Much as with cooking, fermentation makes hard-to-chew food soft. Inasmuch as fermentation often favors the creation of glutamic acid, such foods would have often had umami tastes. Fermentation also destroys some of the bitter tastes in foods and adds complexity to their aromas. Merlin Sheldrake noted as much upon successfully fermenting the apples produced by a descendant of Isaac Newton's apple tree. "To my amazement it was delicious. The bitterness and sourness of the apples had been transformed. The taste was floral and delicate, dry with a gentle fizz. Drunk in larger quantities, it elicited elation and a light euphoria." Sheldrake called the cider "gravity," which seems fitting.

Our ancestors would have chosen to ferment fruits or roots because the result was pleasurable. But they would have benefited from doing so because the result was also nutritious. Fermentation makes calories more available (just as cooking does). It also adds some nutrients to food, including vitamin B_{12} but also, in some cases, nitrogen.[162] Finally, once humans began to rely more on sites to which they returned, fermentation became a way of storing food. Once acidic or alcoholic, fermented fruits and vegetables have the potential to be stored for months or even years. These stored foods could be eaten during the leanest of seasons, whether the dry season in the tropics or the winter in colder realms.

Taste was a field guide employed in the first forays into lactic fermentation, a way to distinguish dangerous from safe. But the nose helped. So too did the biochemistry of the mind. When our ancestors ingested fruits that were a little alcoholic, it pleased them. Their brains experienced pleasure and, thanks to their nose libraries, came to associate such pleasure with particular aromas, those of alcohols and compounds that tend to occur with alcohol. In summary, thanks to taste and the nose's library, those of our ancestors who chose slightly sour, slightly alcoholic fermented fruits would have been both pleased and, perhaps, more likely than their relatives to survive.[6]

Our guess is that Katie Amato is right about the timing: that intentional fermentation of fruits and roots began with ancient humans out in the savanna, where a tipple of sour brew would have been rewarding. If ancient humans could make complex stone tools, it seems likely they could also figure out how to put fruit or roots in a gourd and wait a few days. But Amato's idea is just one model of how the reliance of humans on fermentation began. The other model is even simpler. It requires only a body of water, leftover meat, stone tools, a couple of big rocks, and, once again, a fondness for sour tastes.

In 1989 the paleontologist Daniel Fisher was at an exciting moment in his career. He'd spent a decade studying the remains of mammoths in the far north of North America and Europe and was writing novel papers about the lives of the prehistoric peoples of the Americas. He was an associate professor in the Department of Geological Sciences at the University of Michigan and an associate curator in the Museum of Paleontology to boot. And yet he was also bothered by a set of observations he had made, bothered in a way that keeps a curious person awake at night. The observations related to Clovis archaeological sites in the Great Lakes region of North America. Fisher had studied several sites at which American mastodon skeletons were found in lakes or ponds. In each of those sites, the mastodons showed evidence of butchery but, in addition, were associated with a strange feature. Beside their bones Fisher found gravel and stones that appeared to be positioned as though they had been stuffed inside the mastodon intestines.[163] In one such pond, Fisher even found evidence of a vertical post, as if to mark the spot, "Here rests the mastodon." Fisher was initially puzzled by these finds but would come to believe that they were evidence of situations in which Clovis hunter-gatherers had stored mastodons underwater for the winter, having held them in place by the intestines, which, once filled with gravel and stones, acted like an anchor.

It is easy to imagine the value of being able to store and ferment meat, akin to being able to ferment fruit and vegetable and yet potentially even more consequential. If a small group of people killed a mastodon, or even just a big horse, there would be too much meat to eat in one (or even many) sittings. The meat would then have to be moved when the people moved. As Fisher put it, "Let's say that you manage to kill a mastodon or mammoth. . . . There's no way you can deal with thousands of pounds of meat in an afternoon or even a week. What are you going to do with all of it? Make jerky? If so, will you carry the gigantic load on your back, when dire wolves the size of bears and bears the size of rhinos would be happy to get at the meat?"[164] If that meat could be stored, long-term, and returned to, it would reduce the need to go hunting again right away. It would also reduce the need to fight off hungry wolves and the bears with a taste for mastodon.

But for as much as this advantageousness seemed straightforward to posit, it might have been hard to achieve. It was unclear to Fisher whether Clovis hunter-gatherers (or earlier hunter-gatherers for that matter) would be able to figure out ways to successfully ferment and store meat. Specifically, they needed to figure out ways to ferment and store meat that didn't poison and kill them. A group of bacteria strains of the species *Clostridium botulinum* and *Clostridium perfringens* can readily turn fresh, safe meat into deadly meat as decay proceeds. It has been argued that these bacteria produce toxins as they consume flesh in order to compete with scavengers. Even carrion-feeding animals that have evolved to take advantage of well-rotted flesh are not immune to these microbes. In one study, 90 percent of turkey vultures, 42 percent of American crows, 25 percent of coyotes, and 17 percent of Norway rats that were sampled had developed antibodies to *Clostridium botulinum*.[165] These carrion-feeding vertebrates were all exposed to enough of the pathogen that their immune systems judged it to be a threat, a threat worth remembering.[166]

Could prehistoric Clovis hunters really figure out ways to store meat that would prevent dangerous bacteria from killing them? It is one thing to ferment a fruit by leaving it out, quite another to store and ferment a mastodon or other large mammal and do so in a way that is both safe and appetizing. Fisher decided to do some experiments. He started small. He dropped deer heads in ponds and also into acidic, sphagnum bogs around southeastern Michigan in the fall of 1989 (later, he would do the same with lamb legs). He then came back repeatedly to sample the results, a habit described as "Bambi bobbing" by his students and colleagues. After a month in the pond muck or sphagnum, the uncooked deer heads seemed safe to eat. They were covered in a layer of slime, some kind of microbial growth. But beneath the slime the meat was pink; it smelled of stinky cheese. The smell was not off-putting, but instead almost enticing; it was, Fisher told us in an email, "like blue cheese," a strong one such as Stilton or Cabrales. Fisher didn't know that it was safe, but its aromas, aromas associated with foods that tend to have sour tastes, suggested to him that it might be. As time passed and his confidence grew, Fisher decided to try a more ambitious experiment.[7]

Fisher "borrowed" a 1500-pound draft horse. The draft horse had died of natural causes and was fresh. What was more, the horse was considerate enough to die at a good time of year, in early winter. In late fall and early winter, prehistoric hunters and gatherers would have needed, like squirrels storing nuts, to hoard any food they could for the months of greatest scarcity. And in fact, one of the mastodons found in a pond alongside rocks showed evidence of having been deposited in early winter.[167] Fisher set about butchering the horse. To do so, he first made replicas of prehistoric tools. He then employed them to skin the horse and disarticulate its body. His plan was to submerge the whole horse in a pond, but in pieces, the way one might store meat in the freezer. However, he needed to resolve several challenges, one of which was how to hold the meat down and keep it from floating. Fisher decided to do what the Clovis people appear to have done: he filled pieces of the intestines of the horse with gravel and stones and used them like anchors. Then, carefully, he cut a hole in the ice that had, by then, formed on the pond, and dropped the pieces of horse, some of them quite large, into the lake. It all took time, but not so very much time, and, technologically, the work was easy. At the very least it was no harder that the challenging task paleolithic hunters would have faced in killing the horse in the first place. Over the next months he planned to return to the spot, cut a new hole in the ice, haul pieces of the horse back up, and both sample them for future microbial analysis and smell them.

Fisher waited two weeks before his first sampling. He then walked out onto the ice again and hauled out a bit of horse meat. He used his nose to judge the status of the meat. He sniffed it. It smelled fine, fresh even. It was still edible. He fed it to a friend's three German shepherd-wolf hybrids, and they did not get sick. He also tried a little himself. Then another two weeks passed. It was February. The lake was warming a little. Fisher sampled the meat again. It had begun to smell sour and cheesy, just as had the lamb legs and deer heads. The bacterial counts, later analyses would show, were high, but the aroma of the meat suggested the bacteria were lactic acid bacteria, such that the meat should still be safe. (Here Fisher was judging the meat's condition in just the way Clovis hunters might have, using his nose as a tool.) Fisher decided to build a fire on the ice; he'd seen evidence of such winter, ice fires at Clovis sites at which he had worked. He then took a piece of the horse, which by then smelled like blue cheese, stabbed it on a stick, and held it over the flames. It cooked slowly, too slowly it seemed. Fisher decided to cook it a different way. He waited for the fire to burn down. He then took some of the meat, a thick slice, and put it right on the coals. This worked better. Once the meat was cooked to his satisfaction, medium rare, Fisher began to eat it. It tasted like beef, but sweeter and a little sour.

The study might have ended there, but Fisher was nothing if not persistent. In April, Fisher sampled the meat again. It had become covered in algae, but beneath the algae it was still edible. Its aroma was perhaps even a little more pleasing than it had been, though still very strong. In June, he tried again. The meat was even cheesier and yet still edible. At some point Fisher had a laboratory compare the bacteria on some of the long-fermented lake meat to meat he had kept in his freezer at home. The meat he had kept in his freezer had more cells of problematic bacteria species than did the meat in the lake. The meat in the lake, meanwhile, was dominated by *Lactobacillus*, lactic acid–producing bacteria, lactic acid that may have helped to keep the meat free of pathogens until the spring. By the spring, prehistoric hunters would have been able to hunt again, with ease, and gather too. The stored meat would have become less important. It could probably be abandoned at that point. But just to see what would happen, Fisher continued his experiment. He checked the meat in July and then, finally, August. By August, the meat had begun to fall apart and could no longer be gathered. Even then, however, some seven months after the horse was dropped into the pond, the meat was still apparently safe to eat, just too atomized to collect. As University of Michigan anthropologist John Speth points out in writing about this experiment, had the meat been fermented in a pit rather than a pond, it might have continued to be both edible and gatherable for months more.[168] Fisher had shown that from a single mastodon a family might have been able to eat for no fewer than seven months and perhaps more like eight.[8]

With his experiments, Fisher did not prove that prehistoric Clovis people fermented meat in ponds. Nor did he show (or even suggest for that matter) that earlier paleolithic humans or other hominins did the same. But he did show that it would have been straightforward for humans or other hominins to ferment meat. Very straightforward. In their earliest incarnations (whenever they occurred), such fermentations would have been a kind of clumsy dance in which paleo-chefs interacted with and attempted to anticipate the unnamable powers of invisible (but nonetheless perceptible) beings. Fermenting a mastodon, mammoth, or a horse so that it remains edible and is not deadly appears to be less challenging than making fire or cooking. Maybe Fisher got a little lucky with some details of his fermentation, but some Clovis hunters would have gotten lucky too, and when they did, they could repeat what had worked. Fermenting a mastodon so as to achieve the preferred tastes, aromas, and textures would have required more skill, skill at anticipating the responses of the invisible partner. But there was plenty of time to develop such skill. While Fisher had just one chance to experiment with how best to store meat underwater (his friend didn't have any more dead horses), paleolithic peoples would have had many thousands of chances, over hundreds of thousands of years. Prehistoric humans and subsequent hunter-gatherers could have worked with the microbes around them to ferment meat long before they farmed. If Fisher is right, they did so not just by slowing rot, but also by favoring particular microbes, the microbes that make meat sour, the same *Lactobacillus* bacteria found in fermented fruits. Once Fisher's meat had begun to really rot, he judged its safety by the presence of aromas he associated with sourness and then, later, actual sour taste. Ancient humans, with their ancient and yet fully modern noses and tongues, could have done the same.[9]

Fisher explored one form of fermentation, a form likely to have been undertaken with big animals in Michigan, a state in a region with cold winters and warm summers. Many hunter-gatherers living in similar environments rely or relied on fermentation for winter storage. Once humans began to ferment meat, they could do so in different ways depending on the type of meat or the conditions. All things equal, they would have often chosen more flavorful recipes over less flavorful recipes. They might have been particularly likely to do this in regions where the diversity of other flavors was limited.

Where conditions were dry and the animals being killed relatively small and relatively few, meat could be stored by drying.

Drying requires, somewhat obviously, dry conditions. Hyenas that live on the beach in South Africa dry some of their prey. Leopards hang their kill in trees where it is, compared to the conditions on the ground, drier. Ancient humans might have done the same in the dry season in hot savanna environments. They could have done so even before the advent of fire.

Once they had invented fire, humans could also smoke meat. Smoking is a way of fast-forwarding drying. Our fire-loving ancestors could smoke small pieces of meat, even when conditions were not totally dry, so long as wood was in ample supply. Such smoked meats are the progenitors of modern, unsalted, smoked meats such as the German *bauernschinken*, farmer's ham, which is traditionally air-dried and then smoked using juniper wood. Prehistoric peoples might also have chosen specific woods, with unique aromas, for smoking.

Drying required dry conditions, smoking required fire and firewood. Salting was the third way of making meat dry, but it required, of course, salt. And lots of it. We don't know how wild meats were first salted, but the occasion might have been similar to a description offered by Cato the Elder of the salting of ham in southern Italy. In his work *De agri cultura* ("On Agriculture" or, more literally, "On the Culture of Fields"), Cato writes that

> Hams should be salted as follows; in a vat, or in a big pot. After buying legs of pork cut off the hooves. Use half a modius [about 9 liters] of ground Roman salt per each ham. Put some salt on the bottom of the vat or pot, and place the ham on it skin downwards and cover it with salt. Then, place the next ham on it skin downwards and cover it with salt in the same way. Take care that the meat does not touch. . . . After all the hams are placed in this way, cover them with salt so that the meat cannot be seen; make the surface of the salt

smooth. After the hams stay for five days in the salt, remove them all, each with its own salt. Those that were on top should be placed on the bottom and covered with salt as previously. After twelve days together take the hams out; remove the salt, hang them in a draught and cure them for two days. On the third day, take the hams down and clean them with a sponge, smear them with olive oil mixed with vinegar and hang them in the building where you keep the meat.[169]

Salting of the sort that Cato describes produces delicious meat, full of umami tastes and flavors that result from the slow-motion Maillard reaction that occurs during long fermentations. But it is time-intensive and, more importantly, salt-intensive. Once it was discovered that salt could be used to cure meat and slow fermentation (and favor benign, salt-loving microbes, so-called halophiles), salt eventually became expensive. In most parts of the world, salt curing is a recent invention. It is a delicious, late, ultimately expensive exception rather than the norm, and yet one that would have great consequences for shipping, trade, and the history of Europe, as Mark Kurlansky points out in his book *Salt*.[170] Many salted meats remain popular, including the famous jamón ibérico of Spain, which involves a fermentation similar to the one described by Cato except that it is stretched out over months or even years.

Other types of fermentations would have been variations on the approach Daniel Fisher imagined for the submerged mastodons, wet fermentation. These fermentations might or might not include salt, but shared the characteristic of occurring in liquid. For example, in many regions on the coast, fish were gathered in large quantities. At Norje Sunnansund, an archaeological site on the southeastern Swedish coast, the archaeologist Adam Boethius unearthed thousands and thousands of bones, a quantity of bones corresponding to roughly sixty thousand tons of fish.[171] Norje Sunnansund appears to have been a long-term settlement. People settled there thousands of years before the farming of plants or animals. At the settlement, during the spring, summer, and fall, fish were gathered. Some of the fish were no doubt eaten fresh, along with seal meat, roe deer, wild cherry, sour cherry, and sloe berries. But most were fermented in a dedicated facility. The facility was large and oblong. Postholes mark the poles used to support a roof. Other holes mark where wild bear and seal skins, containing the fermenting fish, were staked. The system was sophisticated. But because salt had not arrived in Scandinavia, this sophisticated fermentation system relied on fermentations akin to the horse in the pond, fermentations in which the animal being fermented would, over months or even years, be fully transformed. It was meat "cooked" by time and microbes until rotten.[10]

The wet fermentation of animals may seem unusual, but it remains extraordinarily common globally. Fermented plants, meats, and fish are important to many indigenous communities of the far north. The Yupik, for example, make a *kuviikaq* by aging plants inside a bag made of the stomach of an animal. The Chuckchi ferment deer blood, deer liver, deer hooves, and the roasted lips of a deer in a deer stomach, along with sweet root. Meat and fat can also be stuffed inside walrus skin to make *tuugtaq*, or roulade.[172] Meanwhile, each modern Scandinavian culture has its own special fermented fish dish (descendants, no doubt, of ancient fermentations like those that occurred at Norje Sunnansund). One of the modern fermented fish dishes loved today in Sweden, for example, is *surströmming*. Surströmming consists of well-rotted herring. It is eaten with *tunnebrød* (thin bread).[11] To those who eat it, it is a delicacy. And yet it has an orthonasal smell that is sufficiently displeasing even to those who love it that it is typically eaten outdoors. The aromas include chemicals that smell like rotten eggs (hydrogen sulfide), rancid butter (butyric acid), and vinegar (acetic acid).[12] Across much of Scandinavia, the old pleasures of fermented fish (and in some places meat) persist. And in much of the world, sauces made of fermented fish are everyday staples. Millions of gallons of fermented fish are consumed annually in the Philippines, Thailand, and Vietnam alone.[13]

Wet fermented meats and fishes can be rich in umami tastes and complex flavors (it is for this reason that fish sauces are so very popular). However, as exemplified by surströmming, such foods often make the difference between orthonasal and retronasal aromas very clear. One can dislike the orthonasal aromas of fermented meat or fish, but love their retronasal aromas and flavors. Mary Tyone, for instance, a Native American woman living in Alaska, said of her people's experience that "when we fix salmon head we put it in a bucket in the ground [for ten days and then] we take it out and eat it." The result is stinkfish. She continues, "Stinkfish, oooh, I love that stinkfish. Smell funny, but it sure tastes good. . . ."[14] In Chinese, *hsiang* refers to dishes that give pleasure by their (orthonasal) smell as well as by their flavor (including their retronasal aromas). Cooked chicken fat, roasted meats, and sautéed onions all have *hsiang*.[173] We need a new word for dishes that have wonderful flavors, but less easily loved orthonasal aromas.

As for how humans came to enjoy different fermented meats, in the abstract it seems possible that the ability to farm microbes on meat and fish was made possible thanks to the human preference for sour tastes. In this, the preference for fermented meats might be akin to the use of microbes to produce sour or alcoholic fruits and roots. But if you get up close and personal with many of the meats or fish that are fermented without salt, smoke, or drying, you will soon notice that, just as for spices, learning to make and love fermented meats and fish requires the ability to learn to love aromas that might otherwise be off-putting. To the extent that any aromas are truly innately repulsive to humans, some of the aromas associated with fermented meats seem like good candidates.[174] In this way, fermented meats and fish are foods that employ (require, really) our whole sensory system. To love a long-fermented bit of unsalted herring is to employ your nose's ability to learn to love nearly anything, your tongue's fondness for umami and sour, and the conscious mind's ability to learn the techniques that yield, repeatedly, the whole set of flavors that the whole system of mouth, nose, and mind love.

Pulling the stories of fermented fruits and roots, meat and fish back together, we can reimagine the history of ancient humans, recent humans, and fermentation. At some point our ancestors began to ferment fruits and, later, roots. It began simply. Taste and aroma guided our ancestors to seek out and make ever more fermented fruits and roots. Sour fruits and roots were safe fruits and roots. Fermented roots and especially fruits also offered the potential for new pleasures, the pleasures of soft textures, sweetness, sourness, and a little buzz (at least after our alcohol dehydrogenase gene evolved its new supercharged form). Perhaps at the same time, perhaps earlier, perhaps later, our ancestors also began to make very similar discoveries about meat and fish. Meat and fish, like fruit and roots, could be left out in a way that made them tastier (more umami), gave them new retronasal aromas and hence flavors and, simultaneously, made them safe to eat for months or even years. Such fermented meat and fish would have been especially important once our ancestors began to kill many fish or big mammals and, in doing so, to have more meat than they could eat at once. Whenever that might have been, the way in which our ancestors would have figured out whether such meat was safe was via its smell but also its sourness. Our tongues then were our way of checking on our ferments which were, whenever they happened, our first gardens: gardens of microbes.[15]

CHAPTER 8

The Art of Cheese

Cheese is milk that has grown up . . . it is preeminently the food of [humans]—the older it grows the more [human] it becomes, and in the last stages of senility it almost requires a room to itself.

—EDWARD BUNYARD, THE EPICURE'S COMPANION[1]

On that day I swore to them to bring them . . . into a land that I had searched out for them, a land flowing with milk and honey.

—EZEKIEL 20:5

Several years ago, we traveled with our kids and our friend Jose Bruno-Bárcena and his family to Jose's hometown of Carreña in Asturias, Spain. Upon our arrival we would meet his extended family. That included his mother, brother, cousin, and second cousin (along with much of the rest of the town). It also included the Cabrales cheese. The cheese is part of the family, a highly esteemed member. In 2019, a single, perfect twelve-inch-wide white and blue wheel of Cabrales cheese sold for 20,050 euros. The cheese was made just five kilometers uphill from Carreña.

The town of Carreña is close enough to the Dordogne region of France that we'd come by car. To get from the art caves in the Dordogne to those in Asturias and the neighboring region of Cantabria, drive toward Bordeaux, then turn south to travel along the Bay of Biscay to the border with Basque Country. At the border, head west. Keep the bay on your right. Stop at the Guggenheim Museum Bilbao, wind your way through the Basque mountains, nibble on some Basque sheep cheese, then not long afterward you will find yourself in among the caves. Visit the ancient art caves of Cantabria, similar to those of the Dordogne and yet wonderfully unique. Visit as many as you can and then, when your kids will bear no more, head for Carreña and the cheese.

When we arrived in Carreña we went straight to the communal cheese cave long used by Jose's family. It is the same cave in which Jose and his cousin Manolo learned how to make cheese from Manolo's mother.[2] As for the cheese cave itself, it may have been used by paleolithic people; it was used by medieval miners; and it served as a family refuge from aerial bombing during the Spanish Civil War. The cave is one of the places where Cabrales cheese comes to life.

The cave is part of a culinary ecosystem. On its roof, spiderwebs hang from stalactites. The spiders eat the flies that might, otherwise, lay their eggs in the cheese. Thick ropes of *Penicillium* fungus grow on everything. Amidst this underworld realm live the cheeses in various states of turning from white to blue with speckling of orangey red. Hundreds of aromas rise up from the cheeses during their transformations, aromas so strong that our son, who loves caves the way other kids love video games, took one whiff and decided that he would sit alone outside.

The paleolithic cave paintings we saw on the way to Carreña required great sacrifice by the painters. The paintings took time.

They required crawling out of the artists' daily lives and deep
into the Earth where the oxygen could be so scarce as to trigger,
as paleoanthropologist Ran Barkai has argued, a kind of mad-
ness. Cabrales cheese is similarly demanding. It always has
been. It requires of those who make it that they shape their lives
not around school plays, sporting events or parties, but instead
around the cheese. In the lives of the cheesemakers of Carreña,
the cheese is a pleasure that emerges out of hardship.

Cabrales cheese requires the milk of goats as well as of sheep
and cows. One needs different equipment to tend to each kind
of animal. Bells to keep track of the sheep. Different bells to
keep track of the goats. Far bigger bells for the cows. And a dog
to keep track of all three. (Generally speaking, the dog does not
get a bell, but one of Jose's cousins, the bell maker in town, has
gotten carried away, and each of his dogs has a different, special,
bell, as do each of his chickens. Such things happen in small
towns.) All of this is made more difficult because the entire system must move twice a year. The goats, sheep, and cows graze
in the valley in the winter, and graze in the highlands in the
summer. It is only when the animals are grazing in the communal highlands and eating the wild, deep green, fat-leafed forage of the high mountains that they produce the milk that yields
the best cheese.[3] In addition, every day that the cheese maker
wants to make cheese, each of these animals must be milked,
typically twice in that day. They must be milked on the same
day even though they are almost inevitably foraging in different
places, over here, over there, their bells ringing from a half
dozen hills. Traditionally, hundreds of miles might be walked
in a year following these animals. The walking is lonely. Often,
the only words spoken would be those offered to the animals
and, Carreña being both a wonderful place and a hard place,
those words might all be expletives.

Once the milk is gathered, it must then be combined. Once combined, it needs to be coagulated, cut, and salted. Then, once it has been salted it must be (at least traditionally) carried back down the hill to town and put into a cheese cave, where it continues to ferment and age. Each and every one of these steps can go wrong. The cheese is especially susceptible to problems once it is taken to the cave. When one cheese goes bad, it can turn others bad.

All of this, beginning with the moment the cow bites a blade of grass with its big dull teeth and ending with the moment you place a piece of cheese in your mouth, takes about two months. When those months are successful, the result is a living ecosystem that is not quite an animal, nor quite a vegetable, an ecosystem with a flavor that is sufficiently powerful so as to be a meal. The result is a living, ever-changing cheese that the people of the valley in which Carreña sits most love. They love it alongside their other traditional foods. They love it with *fabada*, a soup with fava beans, two kinds of sausage, and a base flavored with fatback (what it sounds like, the fat from the back of the pig). They love it accompanied by cider made from three kinds of apples. And they love it in a kind of sandwich, a *cachopo*, made from two pieces of veal between which is inserted, of course, Cabrales cheese. They even love it melted over fries. At dinner time in Carreña, the smell of Cabrales cheese pours into the street through open windows out of nearly every kitchen. It spills, too, out of the cheese cave. It even comes, for reasons we never totally understood, down rivers and streams.

Cabrales is consistently ranked among the best cheeses in the world because of its flavors, complexities, and the work required to yield them. Yet it is hard to say why it exists. Why did a people in a small valley in northern Spain, a valley in which life has, at most points during the last thousand years, been hard and humble, decide to make such a cheese in the first place? The simple answer is that the cheese is a way to store milk for consumption during the hard months when the cows, sheep, and goats are not nursing (and so not producing milk). In this, the cheese is not so different from fermented fish. Cheese buffered hard times. It was a dietary necessity.

What was not a necessity was that the cheese be made in such a difficult way, using three different types of animals, and relying on such a relatively long fermentation of a relatively soft cheese. There is a far easier way to make cheese in Asturias. At the stage at which the curds are formed, those curds could be salted and pressed hard in a form before being allowed to ferment. They might even be smoked to make them even drier. If they were, the cheese would become a hard or semi-hard cheese. Such cheeses are much easier to make. They are easier to ship. They are easier to store. They have long been made that way one valley over. It is probably impossible to know why the people of Cabrales chose such a difficult cheese, but one night while we were having dinner in town with Jose and his family and enjoying cider, Jose offered a hypothesis. "The reason," he said, "is that it is delicious. It is the best-tasting cheese in the world." This deliciousness was heightened, Jose continued, by the humbleness of the other foods available in the local diet. "We used to have to collect chestnuts as kids as our dinner. That was a whole dinner, roasted nuts." In other words, the people of Carreña created their cheese, despite the hardship and occasional hunger associated with that creation, at least in part because it tasted good to them. It tasted very good to them, especially relative to what else was available. We like Jose's idea, perhaps unsurprisingly. It is really just the same idea we have been advocating throughout the book, namely that even when it is harder to find or make flavorful foods, humans and other

FIGURE 8.1. Cabrales cheese cave in Carreña, Spain.

animals sometimes do so. Jose simply extended the idea to agricultural peoples, an extension that hardly seems radical. It isn't radical. But it is hard to test.

What one would ideally want, in order to test this idea, is a case in which many different peoples were simultaneously subjected to a cultural change that led their diets to become less flavorful. It would be great if the experiment could be done on cheeses, because it is relatively easy to characterize cheeses in terms of the strength of their flavors and aromas and their difficulty of production (and the two features tend to match). Fortunately, just such an experiment was conducted at the scale of all of Europe. It was carried out by Benedictine monks.

Beginning in the third century, some Christians in Egypt and Syria began to adopt solitary lifestyles as hermits in the belief that their hardships would focus their longings and draw them closer to their God. These hermits were called *monos* (for "one," in reference to their solitariness) or *monakhos* in Greek and later, in English, monks. Though these monks lived solitary lives, they gathered for religious services. With time, some of the monks began to live together at sites that would come to be called monasteries. But living together proved tricky. There were many practical decisions to be made and upon which one had to seek agreement. For example, which pleasures needed to be renounced? Also, how much praying was really necessary? And what clothes should be worn? There needed to be a rule book. Over time, a number of competing rule books emerged.[4] One of the ones that stuck was inked in 534 by Saint Benedict. It was strict enough to satisfy the very devout (initially) and lenient enough to be popular.

Benedict was from Nursia, a region of Italy known for delicious ham, richly aromatic olive oil, roast doves (with black olives and red wine), and intoxicating drink. He was well aware that a lifestyle that allowed for no pleasurable foods or drinks would be unsustainable, even for very devout monks. He prescribed a diet that was reasonably abstemious and yet not entirely devoid of pleasures. Meals could be eaten twice a day, and at each meal there could be two different kinds of hot foods. Also, he allowed monks a *hemina* of wine a day, about half a pint, unless the weather was hot, or the monk wasn't feeling great, or the work in the field was really arduous (in which case more could be consumed). And, while the monks should not eat the meat of four-footed animals, they could consume their milk. They could also make and eat cheese. With these rules, the monks created for themselves a situation not unlike that faced by the humble cheesemakers of Carreña. The flavors in their diets were more satisfying than those of more ascetic monks and yet scarce relative to those they desired. The food that they could work with to create something better was primarily milk.

The question, per Jose's hypothesis, is whether, when confronted with a relative dearth of flavor, the monks would make cheeses in which flavor and aroma were featured. In this way, the monastery was a culinary experiment. More than that, it was a replicated experiment. Monasteries in different regions made relatively independent choices about the foods they made and did so in the context of very different local cultures, languages, and climates. Each monastery was a partially independent test of Jose's hypothesis. If the monks were to choose to focus on making flavorful cheeses, they had several things in their favor. Monk culture, like the culture of Carreña, valued hard, patient work. The Benedictine motto was "To labor is to pray." Conversely, idleness was the "enemy of the soul." To avoid idleness by laboring among the wheels and wedges of cheeses, then, was, in and of itself, godly. And because the monks took large amounts of land and were given even more (often by those interested in securing a good spot in the afterlife), it was godly work that could be carried out at a large scale. As a result of the patience of their piousness and the luxury of land, monks innovated new approaches to agriculture, created new foods, and took credit for and helped to preserve the techniques and ingredients necessary to make old ones, or at least the subset of old ones that suited their needs and preferences. In this, their relationship to ancient foods was akin to their relationship to ancient documents. The monks copied and translated by hand the subset of the classical literature and science of ancient Rome and Greece that they believed to be most important. So, too, they translated the ancient foods of the places in which they lived, again by hand, passing on to future generations the versions most salient to their own lives.

Like the cheesemakers of Carreña, the monks had choices with regard to the kinds of cheeses they would translate or innovate; their options fell into three basic categories: fresh cheeses, aged hard cheeses, and aged soft cheeses. In ancient Rome, before the time of Saint Benedict, fresh cheeses and aged hard cheeses appear to have been the main types of cheese. The process of making either of these types of cheese starts in the same way. The cheese must first be "curdled," which is to say that the curds and whey in the milk must be separated. In making fresh cheeses, this is mostly the end of the story. Fresh cheeses that rely primarily on curdling include modern cream cheese and chevre. In such cheeses one can taste the influence of the food of the animal that made the milk. Fresh cheese from animals fed hay are mild. Those from animals that have walked up and down hills, being followed by and following shepherds, are different in each place they are made. In fresh cheeses from animals that wander one can note the flavor of the particular

FIGURE 8.2. St. Benedict of Nursia handing off the rule book. While in general Benedict eschewed extravagance, obviously the artist who depicted him here thought he wouldn't mind a fabulous chair.

grass species, or the differences between the flavors of cow cheeses and those made from, say, goats. Goat milk and cheese have fatty acids that are not found in cow milk (such as 4-methyl-octanoic acid). Buffalo milk cheese has still different fatty acids, some of which smell like mushrooms. Fresh cheeses are made nearly everywhere cheese is made, but they are the cheese of the day: they don't store well. This is particularly true in southern Europe, where warm temperatures cause the fresh cheeses to go bad very, very quickly. Aged, hard cheeses were the ancient solution to shipping and storing cheese.

The making of aged hard cheeses requires more steps than are necessary for fresh cheeses. The curd must be shaped; often it is cut into pieces and then scooped by hand[5] into a form of some sort. The curds are then pressed to draw their water out and, sometimes, also salted (which draws even more water out). The result is a hard cheese with enough water left in it to allow a predictable set of drought-tolerant bacteria and fungi to colonize, but not so much that the cheese goes bad. Those drought-tolerant microbes metabolize the easy-to-eat proteins and fats and leave nothing left over for other species. In their metabolic work these bacteria and fungi simultaneously imbue the cheese with new aromas and hence flavors and prevent more problematic microbes from taking over. The most ancient of these hard and flavorful cheeses included cheeses similar to Parmigiano-Reggiano cheese (aka parmesan), Manchego, Asiago, and Gouda. Such cheeses can be very different from each other[6] and yet are produced using similar processes and are united in their durability, their hardness. The first evidence of these cheeses comes from metal cheese graters found in southern Italy (around what is now Naples) as early as 700 BCE. Southern Italians have been shredding parmesan-like cheese on their food for no fewer than 2700 years.

In making these two kinds of cheeses, ancient cheesemakers undoubtedly stumbled upon the third major type of cheese, aged soft cheese. But as far as history records, any aged soft cheeses that resulted from such stumbles appear to have either been very regional or short-lived. They are not noted by ancient Romans historians. If they were being made anywhere, they were being made in places upon which the Romans chose not to comment. There is a reason that such cheeses might have been avoided. Compared to aged hard cheeses, such as parmesan cheeses, aged soft cheeses are demanding and dangerous. A hard cheese is akin to a cured meat.[7] The focus is on favoring particular microbes by altering their environment. Making soft and aged soft aged cheeses is a kind of slightly out-of-control waltz with an invisible but increasingly strong-smelling partner. One goes through practiced steps, hoping that they yield the desired outcome, but never fully sure about the quality of the results until the music stops (and the cheese is cut). What is more, a waltz with aged soft cheese can be dangerous. The cheese can go bad, but it can also be taken over by pathogens. Such cheeses would have been particularly hard to make successfully in southern Europe, where the warmth and moisture in the air would have, as cheese historian Paul Kindstedt puts it, "rendered the cheese inedible by microbial spoilage" within a few days. However, as Kindstedt goes on to note, "in the cooler, damper climate of northwestern Europe very different outcomes were possible when the right conditions were satisfied."[175] Yet, even in northwestern Europe it was not trivial to meet such conditions; and doing so was unnecessary wherever and whenever an aged hard cheese could be made.

However, aged soft cheeses did offer an advantage: extraordinary flavors. They have a mouthfeel that can be very meaty. They have exceedingly high concentrations of compounds that trigger umami taste. And, they tend to have aromas that smell bodily. These aromas smell bodily because such cheeses, in their body-like wetness, favor microbes more often found on bodies, including human bodies. And the older the aged soft cheeses grow, to paraphrase the writer Edward Bunyard, the more human they become.

If the monks tended to seek out stronger flavors, especially flavors more reminiscent of meats, regardless of what was easiest or optimal, it is these aged soft cheeses that we would expect them to have made. Conversely, a monk interested only in making cheese that will store, ship, or be a satisfactory contribution to the diet should never make a semi-soft cheese. There is no need. Lumps of soft, fresh cheese and wheels of hard, old cheese might easily have been the extent of the story of monks and cheeses. There is no official tally of the most aromatic or meaty cheeses of the world. Nor has anyone yet statistically compared the recipes or chemical composition of aromas of cheeses associated with monasteries with those with other origins. It would be possible, though time consuming. It is the kind of project on which a graduate student in monastic history and a graduate student in microbiology might collaborate while traveling joyously across Europe eating cheeses and reading ancient manuscripts. (If that sounds like the two of us regretting that neither of us has yet learned Latin, well, it is.) Meanwhile, it is clear that many monasteries in France and elsewhere came to specialize on the hardest-to-make and most strongly smelling aged soft cheeses, cheeses that rewarded the monks with lifetimes of mystery. In some cases, the monks may have copied cheeses from peasants. But they also invented many such cheeses anew.

Some of the aged soft cheeses made by the monks relied on quick coagulation (thanks to the use of large amounts of rennet).[8] If these cheeses were allowed to ferment in cool conditions that were dry (or at least not too humid), they favored white and grey species of *Penicillium* fungus, including *Penicillium camemberti*. Once colonized with these fungi, such cheeses become covered with the "bloom" of the fungus and are hence called "bloomy-rind" cheeses. Bloomy-rind cheeses include Brie de Meaux and Camembert (hence the name of the fungus). Bloomy-rind cheeses are also referred to as "surface ripened." Most cheeses become harder as they age, but these bloomy-rind or surface-ripened cheeses become softer. The *Penicillium* fungi ripen the cheese by growing from the rind (which has a white color and a flavor often described as "mushroomy"); as it does, the interior of the cheese slowly liquifies and becomes creamy.

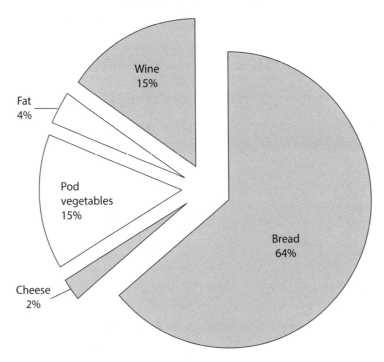

FIGURE 8.3. An example of the Benedictine diet. Daily food and drink consumption (by weight) at the St. Germaine des Prés Monastery in France in the year 829. Grey pie wedges show fermented foods and drinks. Monastic diets were not very seasonal since most of what they ate was fermented and hence could be stored and accessed year-round.

If, on the other hand, very similar cheeses were allowed to ferment in moister caves, such as those in Carreña, another species of *Penicillium* (often *Penicillium roqueforti*) grew and blue (or bleu) cheeses such as Cabrales, Roquefort, and Stilton were the result.[9] Some of these cheeses are intentionally poked with stainless steel needles to create holes through which the *Penicillium* fungi grow. This creates the riddling of blue so characteristic of Roquefort or Stilton. Cabrales is different; as with bloomy-rind cheeses, the *Penicillium* grows from the surface of the Cabrales cheese in. Which of these approaches is better is a great thing to start an argument about while in Carreña or anywhere else in Asturias, especially after a few glasses of cider.

The most unique of the cheeses that the monks copied or invented, however, were neither bloomy-rind cheeses nor blue cheeses. They were instead washed-rind cheeses and smeared-rind cheeses. Let's start with the washed-rind cheeses. These were even harder to produce and even more mammalian in their countenance than bloomy-rind or blue cheeses. Washed-rind cheeses are, like blue cheeses, stored in relatively humid caves and cellars. But, unlike blue cheeses, they are washed with very salty water as they are aging. The salt favors bacteria that prefer dry conditions. The bacteria favored by such washing include the species *Brevibacterium linens* and its relatives. *Brevibacterium linens* is a denizen of both cheeses and the salty, dry habitat of human feet. It grows in orange blotches and patches on the surface of the cheese. The bacteria on washed-rind cheeses also include species from the ocean such as species of *Halomonas* and *Pseudoaltermonas* that appear to be introduced to the cheese with the sea salt.[176] The bacteria favored by washed-rind approaches are finicky. They require oxygen and so only grow on the surface of the cheese. They also don't grow under acidic conditions and so grow only once the fungi on old cheeses have already consumed most of the acid produced by lactic acid bacteria.

Cheesemaker David Asher, author of *The Art of Natural Cheesemaking*,[177] suspects that washed-rind cheeses are unlikely to have been common anywhere before they began to be produced in monasteries. For one, such cheeses require larger quantities of milk, more milk than could be produced in a day by one animal. Monasteries were more likely to have access to larger quantities of milk than were poor farmers. In addition, however, making such cheeses required (and requires) washing the cheeses with salt brine every day, even on holy days. Every single day. It required doing so even though the primary benefit of the effort was with regard to the sensory experience of the cheeses, their tastes, aromas, and mouthfeels.[10] It required the work and care that Saint Benedict had asserted to be so important. It also, however, required a fondness for the culinary pleasures that hard work can yield. Washed- and salted-rind cheeses are meaty, fleshy, and toothsome; in being eaten, they bite back.[11] They bite back with a pleasure that the monks enjoyed as though it were meat, even though they were not allowed to eat meat, perhaps especially because they were not allowed to eat meat.

Nor were the washed-rind cheeses the end of the ways in which the monks figured out how to further manipulate microbes. The monks also discovered that they could take a cloth or other material and spread the orange patches (which were, although they didn't know it, colonies of *Brevibacterium linens*) around the outside of the cheese or even from one cheese to another; in doing so they created wiped-rind cheeses. When these patches were spread, the *Brevibacterium* would grow to coat the entire cheese. As it did, new aromas appeared, aromas associated with the breakdown of proteins, aromas that can be smelled miles from where these cheeses are being made and each and every place they are cut, plated, and eaten.

For the monks, to work hard was to be devout. To find and feature new flavors, during such devotion, was to be human. And to gain control of the factors that would yield one type of set of flavors versus another was science. Meanwhile, to exercise that control in different ways in different monasteries represented some mix of the monks' devotion, humanity, science,

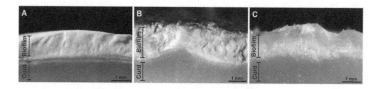

FIGURE 8.4. Cheeses. At left (A), a bloomy-rind cheese on which *Penicillium* fungi have produced a thick layer of life. In the middle (B), a rind like that found on parmesan cheese. At right (C), a topographically complex *Brevibacterium* rind on a washed-rind cheese. In each case one can see both the curd (bottom) and the biofilm of life that grows upon it.

and particular monastic cultures. It was in the context of washed-rind cheeses that devotion, humanity, and science most fully came together.

Washed-rind cheeses were discovered and elaborated independently in monasteries across Europe. In each monastery in which they were made, they were made in a slightly different way, or even more than one way. If the cheese was somewhat dry and wiped for only part of the ripening, a Gruyère could be a made. The monks of the Gruyère monastery in Switzerland would come to specialize on (and find pleasure in) this cheese. Meanwhile, in the French monastery of Maroilles monks figured out a slightly different approach. Maroilles is made using cow's milk that is allowed to ferment with more moisture than for Gruyère; it is square and orange and has an aroma that is at once rotten, meaty, and fruity. Monks at an Alsatian monastery in the Vosges mountains in France tried something different. They brined their cheese again and again and created Munster.[12] Then there is Époisses cheese, made at the monastery of the same name. It is among the least dense and most moist of the washed-rind cheeses. It gets some of its special flavor from the brandy the monks use to wash the cheese (a brandy the monks also make); the rest of the magic, though, is microbial, powerfully microbial. Stéphane Hénaut and Jeni Mitchell, who together wrote *A Bite-Sized History of France*, noted that Époisses has an aroma that threatens marital harmony (which, of course, didn't matter to the monks).[13] Other monks discovered that cheeses could be washed in beers rather than brine, a practice continued at the Chimay and other monasteries today.

To summarize, the monks came to make exactly the kinds of cheeses they would be expected to make if they were seeking out flavor. Whether or not they did so with a greater frequency than cheesemakers outside of monasteries awaits more study (though it seems likely). There were, however, two other elements to this story. One is that the monks sold some of their cheese (in some cases, much of their cheese). When they did, they were making cheese that suited not only their own flavor preferences but also those of wealthy city dwellers (the medieval version of hipster gastronomes). One can imagine that just as with modern art and food, sometimes these patrons influenced the cheese being made, and other times they simply made cheese making economically viable without having a big effect on the monks' approaches to cheese. However, the influence of patrons is likely to have been greater on aged hard cheeses than soft cheeses. As Paul Kindstedt points out in *Cheese and Culture*, aged soft cheeses and, particularly, washed-rind cheeses, would have been hard to transport to market without damage.

The other element relates not just to which cheeses might have most appealed to monks, but more specifically, the ways in which monks might have tended, through the workings of the olfactory system, to learn to love complex cheeses. As Gordon Shepherd describes in his book *Neuroenology*, when people are trained to identify individual aromas, they can often learn them quite well.[178] A group of people in one study, for example, learned to correctly identify seven aromas 82 percent of the time. But when the aromas were mixed in pairs, the ability of the same people to identify one or both of the aromas dropped dramatically (to about 35 percent depending on how one does the math). If three of the seven aromas were offered at once, the participants got them all right only 14 percent of the time, and if four were offered, just 4 percent of the time.[179] Something similar would have occurred with new monks trying to identify the aromas and flavors in the cheeses. However, it is also known that the more time someone spends trying to distinguish different aromas (or even just being exposed to them), the better their ability to distinguish those aromas becomes. In part, groups of aromas come to acquire their own names and identities in the brain. In addition, individual aromas and memories associated with those aromas become more conspicuous in the mind and easier to recall. As a result, the more someone eats a food with very complex aromas, the more that food offers to them in terms of identifiable aromas, and flavors as well. And because foods like cheese, wine, meats, and fruits are so complex (hundreds of aromas, not four) even once someone begins to distinguish the dominant aromas, new aromas appear and can then be learned. As Shepherd put it, "learning to discriminate between two wines primes the olfactory cortex for heightened discrimination between subsequent wines." The same is true of cheeses. For the monks, learning to discriminate good cheeses and bad cheeses, meaty cheeses and less meaty cheeses, "heightened their discrimination" between subsequent cheeses. As they refined their abilities, they also increased their abilities to appreciate and enjoy the differences that they discerned. In this, they were not alone.

This story of the monks and cheese is strange and unique. Yet it is general in that complex flavors and aromas have come to be favored in many cultures and contexts, especially when the flavors available are limited. The rarity of flavorful foods might be part of what makes some (but not other) chimpanzee communities use sticks to find ants. It might be why Clovis hunters chose to hunt some prey rather than others. It may be a component of why ancient Mesopotamians flavored their stews with spices. We were reminded of the repeating nature of the story of flavor recently when Rob was on a trip in Okinawa, Japan. He'd gone to meet with colleagues trying to chart out the global geography of ants, what ants eat and why (at the moment, his scientific life is roughly equal proportions food, primates, and ants). On the last day of the meeting he went to dinner with his former student Benoit Guénard. Benoit is now a professor in Hong Kong but lived for years in Okinawa, where he also met his wife. Benoit's family, then, is Franco-Japanese (though Benoit would be quick to say that he is not really from France but instead Brittany, a region of France with a strongly autonomous spirit). Benoit's family has a great fondness for both French and Japanese foods. Rob was delighted to be able to sit down with Benoit at a traditional restaurant in Okinawa. The two ate various pickles, a local seaweed, peanut tofu and noodles with squid ink, and then, Rob ordered tofuyou.

Tofuyou appears to have been innovated by cooks, including Japanese monks, on the basis of older Chinese traditions, during a period in which the monks were (similar to the Benedictine monks) encouraged to eat vegetarian. Tofuyou is made by taking soybean curds (similar in many ways to cheese curds) and drying them and, in doing so, allowing them to ferment slightly. The curds are then washed in awamori (much as Chimay cheese is washed in beer), a spirit distilled in Okinawa from rice. Washing the tofu with awamori disfavors the many microbe species killed by alcohol and dry conditions. Meanwhile, it favors species tolerant of alcohol and dry conditions. (Other fermented tofus rely on brines in a process akin to brine-washing a cheese.) Finally, the tofuyou is ripened, or as cheese-makers would say, finished.[180] Superficially, fermented tofu and well-aged semi-soft cheeses have many similarities. Both begin with curds. Both require stages of fermentation. And finally, both can employ either salt or alcohol to favor some species over others. What Rob was not prepared for as he brought the tofuyou up to his mouth was that it would feature aromas and flavors similar to aged cheeses. The tofuyou had a soft texture similar to that of, say, Brie, and a range of rich flavors that gave way to a final flavor, an aroma in the back of the mouth, of Cabrales blue cheese.[14] The human story is infinite in its dimensions; culture and learning can lead different peoples to like very different food and yet, when it comes to deliciousness, some stories repeat.

CHAPTER 9

Dinner Makes Us Human

*Food and language are not only close neighbors ... they occupy the
same house.*

— GORDON SHEPHERD

While staying in the small town of Bouriège, in the foothills of
the French Pyrenees, we and our kids were invited to the town
fête.[1] That we were invited said nothing special about us. Everyone in town at the time was invited; all of the townspeople and
all of the visitors regardless of class or condition were asked by
the mayor to join in. Young children came. Old couples came.
Mean people, kind people, funny people, boring people. All the
people. The evening began with informal drinks and conversations. We each had a glass of a locally fermented Blanquette
wine invented by monks nearly five hundred years prior. After
the wine, as the sun set, we sat at one of two long rows of tables
set up on a bluff above a meandering stream. We sat between
the British artist Alvin Booth, who specializes in photographs
of bits and pieces of the bodies of nude dancers, and a shepherd
who still smelled faintly of sheep (an aroma that we can tell you,

thanks to the reading we've done while writing this book, is due in part to the scent of capric acid).

As we talked, we were treated to more wine, then a salad, then the ribs of a pig brought around by the shirtless man who had cooked them, then potatoes thrown to us by a round-bellied chef in a beret pushing a wheelbarrow, and then a fruit dessert handed to us by a group of young girls. Finally, when the night had grown sufficiently dark so that we could see each other primarily in the flicker of the candles, cheese plates appeared. We could see the ancient geometry of the cheeses, squares, triangles, and trapezoids; but it was far easier to smell them, sniff at their differences. Then the music and dancing began. Three musicians, one playing a clarinet, a second a fiddle, and the third an accordion, walked up and down among the tables. More wine was served.[2]

While immersed in the flavors of the wine and cheese and the sounds of the music, we talked about art, history, monks, sheep, the best places to fish in a certain part of England, and who should or should not be Bouriège's next mayor. We also discussed a project Alvin Booth and his wife Nike Lanning had recently completed. They had recorded twenty-four hours of Radio France Culture programming and then isolated the non-word sounds produced by French radio hosts and guests—the oos, the ahhs, the pfffs, and much more. The result (which we later heard played for us) was a kind of soothing, unintelligible music. Then, at some point, before the music but after the cheese, we started to talk about meals among chimpanzees.

It was a more obvious topic to consider than it might seem. We had spent much of the summer in Germany. While there, chimpanzees and their dining habits were a frequent topic of conversation. Rob worked with chimpanzee researchers during the day and then we both met with them for drinks or dinner during the evenings. What was more, by the time we departed Leipzig we'd been left with a surprising set of realizations about dinner parties, sharing, language, and what does and doesn't make humans unique.

The realizations were spurred by a chance encounter. We bumped into Roman Wittig at a backyard party.[3] Roman Wittig spends part of his life in Leipzig, among people who roast pigs and host dinner parties in lovely backyards, and part in the Taï Forest in Côte d'Ivoire, among chimpanzees. He is one of the world's experts on the ways in which chimpanzees eat and share food.

In his time in Taï, Wittig and his students and colleagues have observed chimpanzees for hundreds of thousands of hours. In one recent study, a student, Liran Samuni, working with Wittig and other colleagues, observed two groups of chimpanzees for roughly two thousand hours each. Two thousand hours in the field, watching, taking notes, and recording data. During those hours, Samuni and her team documented 312 independent instances in which chimpanzees shared food, whether the meat of red colobus monkeys, fruit, or seeds. Roughly once every other day at least one of the forty chimpanzees being observed shared food. Through their observations of these events, Samuni, Wittig, and colleagues were able to show that in sharing food the chimpanzees of Taï Forest appear to obey a set of three rules.

The first rule of chimpanzee sharing is that chimpanzees share with individuals with whom they have worked to procure a food item. This is particularly true if it is a hard-to-procure item such as the flesh of a colobus monkey.

The second rule is that, apart from sharing with their fellow procurers, the chimpanzees share with individuals with whom they have developed long-term social bonds, or with whom they would like to develop such bonds. In other words, they share with their friends and potential friends.[4][181]

The third rule is that sharing with friends tends to be stratified by social class. Marginal chimpanzees at Taï do not share with dominant, powerful chimpanzees, but instead with other marginal individuals. Conversely, powerful individuals share food with other powerful individuals.

Wittig and another group of colleagues also studied whether food sharing affected the levels of oxytocin in the chimpanzees' urine. As our friend Heather Patisaul, a neuroscientist, put it, "oxytocin promotes trust and bonding. It does so largely by reducing anxiety. Moms (and in some species, dads) get a surge of it when they become parents." What drives the response, though, is dopamine. Oxytocin production triggers dopamine production, which makes "all that baby cuddling feel good, and makes partners feel good about each other in monogamous, social species."

It has been known for a while that chimpanzees produce extra oxytocin when grooming or being groomed. Wittig and his colleagues were able to show that sharing or receiving food triggers an even larger increase in oxytocin (at least as measured by the concentrations present in urine), particularly if the food is hard to obtain.[182] This was true if the chimpanzees were sharing with friends. It was also true if the chimpanzees were sharing with animals they wanted to be their friends. Collectively, the research of Wittig, Samuni, and collaborators suggests that in sharing food, chimpanzees get an oxytocin-triggered dopamine rush. That pleasing rush reinforces existing social bonds and also helps to establish new bonds.[5] It does so through pleasure. Chimpanzees choose foods with pleasing flavors and share those same foods because doing so can, in some cases, yield pleasure yet again.

Back in France, as we talked to strangers at the dinner party, shared food, and swapped stories, we too felt pleased. The people around us seemed pleased. We were, together, steeping in our mutual oxytocin, oxytocin and wine. It felt easy to imagine our interactions as layered on the ancient rules and biochemistry shared by chimpanzees and our common ancestors. And yet, we were also aware of how much more complex a dinner party can be than is the sharing that occurs between chimpanzees. This complexity is woven, person to person, via conversations, conversations built of spoken words.

Our ability to speak allows us to invite relative strangers to our tables, but also to choose which strangers. Our words bind us socially. It is through words and stories that we relate to each other. Across cultures, food, conversation and negotiation are linked. Chimpanzees bond by handing each other food and grooming each other's fur for lice. We also bond by physically handing each other food, and taking food, but we refine those bonds with our words. Words have replaced fingers as our most common social connection.

It isn't that chimpanzees don't communicate with sounds. They do. But their sounds are holistic; each sound expresses a complete sentiment. The sounds chimpanzees make are also always manipulative. Their sounds are produced with the goal of convincing another animal to do something. For example, Ammie Kalan, a primatologist focused on chimpanzee communication, has shown that when chimpanzees get together in a tree filled with good fruit, they make a call that means "Come get good fruit here." Or at least that is what they do if there is enough fruit to share and if they know that one or more of their friends has been left behind.[183] Ammie can understand these messages. She can tell that there is good food and maybe even how much.[184]

In listening to one another's calls, chimpanzees can even learn to recognize who is calling, such that a call might really mean something more like "Nick says come get good food here." To the extent that pronouncements of chimpanzees contain additional adjectives, they relate to quantity or quality: "Nick says come get REALLY GOOD food here," where the emphasis comes from some combination of pitch, timbre, and volume. The quantity and quality of food can also be conveyed in another way, as a function of the number of chimpanzees that are present. The more good food there is, the more chimpanzees might call, the louder the din becomes, and the more excited the receiver of the call might be. In chimpanzee grammar, the collective din is the exclamation point. The collective din comes from the collective sound of the food-inspired vowel-like utterances. It is a din that is superficially similar to the sounds produced by the French between, before, and after words, the sounds on which Booth and Lanning focused, but multiplied by the size of the party. It is the sound we stumbled upon when walking toward the fête in Bouriège, a kind of song sung with mouths, lips, and vowels, the music of enthusiastic pleasure.

Yet, wild chimpanzees show no evidence of calls that express more complex sentiments or refer to objects (such as fruits) that are not right in front of them. They also do not understand the thoughts of other chimpanzees. Nor do they attempt to manipulate those thoughts. Nor do they invent new calls. Different chimpanzee populations, despite having different culinary traditions, tools, and foods to talk about, say the same things. Dinner for chimpanzees, to the extent that it entails communication, is a never-ending story about the presence of fruit or meat and how good it is. As guests, chimpanzees are enthusiastic bores.

FIGURE 9.1. Chimpanzees in Taï Forest, Ivory Coast, having a small dinner party featuring the dismembered body of a colobus monkey.

Back at our table in Bouriège, as our dinner companions spoke easily in French, then English, then French again, about fish, sheep, food, and neighbors, we enjoyed what was before us. But we also found ourselves wondering how the complex human ability for speech evolved. We talked about this question as Rob grabbed yet another bite of cheese off the common plate, tore off a piece of bread from the common loaf, and asked a stranger to pass the wine. The ability to invent new words and sounds may have arisen among ancient humans during mealtimes. It is during meals that they would have needed to negotiate rules of sharing food more complex than those in which the chimpanzees of Taï or our ancestors might engage. Even if language didn't arise in the context of food, it seems clear that language became more complex in the context of meals. Language allowed meals to entail, as Brillat-Savarin put it, every form of sociability: friendship, love, "business, speculation, power, importunity, patronage, ambition, intrigue." As a result, while the rules of dining vary in many ways among cultures, the importance of eating together transcends human culture and time. And this togetherness improves the deliciousness of food; it improves it through the pleasure of companionship and the pleasure of stories.

For many thousands of years, fireside gatherings were where our ancestors shared stories. It was around fires, with food in hand, that our ancestors communicated their knowledge and understanding. It was in such gatherings that truth could be sorted from falsehoods. Fireside chats were long our system of peer review and consensus building; they were the first universities, scientific societies, kitchens, and dining rooms all in one.

Around fires, our ancestors shared what they knew about how to categorize and use the plants and animals around them. They studied the timing of biological events (and when and why a particular plant or animal was most delicious). They did so in order to survive, but also to make their own lives more pleasurable. This relationship between food and discovery continues around dinner tables and in kitchens around the world. It is also a core part of the Western scientific tradition. Greek scholars gathered at a symposium, a word that describes drinking (*posium*) together (*sym*). Scientific meetings continue in this tradition. Scientists get together over drinks or food to talk about ideas. In addition, individual scientific breakthroughs are very often associated with particular meals or series of meals. Charles Darwin gathered many of the data and observations that he would use to make sense of natural selection on the voyage he took on a ship called the *Beagle*. As the *Beagle* nosed along the shores of South America, Darwin started to see both that species evolved and the role that the survival of the fittest played in that evolution. But Darwin wasn't on board the

Beagle's journey as a scientist. He was invited to come along as a messmate to keep the ship's captain, Robert Fitzroy, company and, in doing so, prevent him from succumbing to the loneliness and despair common on long sea journeys. For this job (which did not pay), Darwin's qualifications were twofold. He was of the same social and intellectual class as Fitzroy (much in the way that chimpanzees share with chimpanzees of a similar social status). In addition, at meals to which he had been invited around England, Darwin had gained a reputation as a good dinner companion. He had interesting things to say. In other words, Darwin was able to make his great discoveries, discoveries that made the workings of the biological world clearer than they had ever been before, because, as Martin Jones put it in his book *Feast,* "he enjoyed the pleasures of the table."[185]

We could argue that it was flavor and deliciousness that first led humans to communal sharing, language, the refinement of language, and even, ultimately, science (and the ability to study flavor and food). But we've gone too far here, and we know it. Rob's friend Nick Gotelli often offers up a quote from his Russian grandmother (or maybe she was Polish; it seems to change from one telling to the next), who said something to the effect of "When you have a new hammer, everything shiny looks like a nail." Flavor is our new hammer, and so we are probably whacking some shiny things here that aren't nails. What we do know is that to whatever extent flavor and food were central to the beginnings of the scientific search for the truth, they aren't any more. It isn't so much that the study of food and flavor has become marginal as that it has become atomized. The shepherd and the artist rarely sit together to talk anymore, nor do the neuroscientist and the food scientist. Food scientists study how to make more of a particular food, or how to improve a particular flavor (often one produced in enormous quantities). Food safety specialists study how to control food-borne pathogens. Ecologists study the interactions among organisms in food, or the relationship of foods to the environments from which they come. Evolutionary biologists study the history of foods and the senses we use to appreciate them. Neuroscientists study the brain's response to individual chemicals. Paleoanthropologists dig for teeth and use them to see as much as might possibly be seen from a bit of ancient dentition. Home cooks carry on traditions of cooking. Each of these scholars and citizens sees some part of the bigger picture, but it is no one's job to step back and unite all of the observations and ideas.

This book gave us an excuse to have many gatherings in which we pulled together people that might not otherwise talk to each other. We were able to stage many fireside chats, whether they were around actual fires, around old stoves in traditional houses in Catalonia, around tables in western France, or at the Max Planck Institute cafeteria. In doing so, we were able to see more of the big picture than we might have otherwise seen. The results are the stories we've shared with you in this book. These are stories we could not have learned and retold without the insights of many people, as well as the connections made among people who don't normally talk (the oyster biologist and the oyster historian, for example, or the chimpanzee researcher and the expert on honeybee honey). But it is also true that most of what we could have written about we haven't.[6] We didn't yet have the right dinner, the right gathering. The pieces still need to be united in order to see the whole. In a way, that is the good news. How boring it would be if all of the exciting discoveries had already been made and all of the interesting conversations had already been had.

There is enough left to study and talk about with regard to flavor and evolution to keep people busy for centuries, probably longer. It is the nature of our species to savor the pleasure of great flavors, but also to savor the pleasure of considering their causes. This search is even embedded in our very name. *Homo sapiens*, the name of our species, is often said to mean "knowing" (*sapiens*) "human" (*Homo*). But *sapiens* originates in a verb meaning "to taste" and later "to have discernment." One might then also read the name of our species as the human (*Homo*) who discerns through taste (*sapiens*) or flavors. We discern and choose through flavors, but we also search, research, and learn by tasting, and are uniquely suited to doing so together with others of our species, whether around a fire or at a table. We sit together and make sense of the world one bite at a time.[7]

NOTES

Prologue: Eco-Evolutionary Gastronomy

1. Michael Tordoff, a specialist in taste at the Monell Sensory Center, upon reading this paragraph, shared a story about a meeting of the Society for the Study of Ingestive Behavior (which is a fancy way of saying "The Society for the Study of Eating"). The members of the society were asked to submit three to ten keywords that described their interests. There were lots of specific, technical keywords, Michael said, such as "brain mechanisms," "cholecystokinin," and "meal patterns." But only one scientist out of the three hundred present said "pleasure." That scientist's answer stood out so much from among the others that, twenty years later, Tordoff still remembered his name.

2. That would have been a little long, though it might have worked as a subtitle. Brillat-Savarin's own subtitle was a wonderful mouthful, "Meditations on Transcendental Gastronomy. A theoretical, historical and topical work dedicated to the gastronomies of Paris by a Professor, member of several literary societies."

3. The word "delicious" comes from the Latin *deliciosus*, which describes something that is pleasant, sensuous, or even voluptuous.

Chapter 1. Tongue-Tied

1. Or as Lucretius himself put it, "The larger observable bodies—the sun and the moon—are made of atoms, just as are human beings and water flies and grains of sand."

2. Historians believe that the manuscript was found in the Monastery of Fulda. Many of the monks of Fulda focused on copying manuscripts. Manuscripts were borrowed to be copied (in order to increase the size of the library), and in addition, manuscripts already in the library were copied so as to keep them from being lost. The library would ultimately come to hold no fewer than 2000 ancient manuscripts.

3. David Norbrook, an emeritus professor of literature at Oxford University, noted to us that for Lucretius there was no hierarchy among species and the kinds

of pleasures they experienced. Humans, mice, and fish were all, first and foremost, *animantum*, living creatures, living creatures guided by pleasures. However, as Norbrook went on to note, Lucretius did believe that the things in which humans found pleasure could be chosen and refined. We are guided, like the other animals, by our senses and their manifestations in our minds, but can also learn to love new things, or old things in new ways, a reality we consider in more detail in chapter 3.

4. A sentiment captured in the Japanese custom of saying *itadakimasu*, thank you, before eating. The thanks are not to the chef but instead to the organisms whose elements are passing into the body of the diner. *Itadakimasu*, thank you for your life.

5. And to add an additional layer to this challenge, species must not only find these elements in the right ratios, they must also find any compounds that their bodies are unable to make on their own, even with these elements. For example, human bodies and those of some other species cannot make vitamin C, even if they have the right ingredients to do so. They must find and eat organisms that contain vitamin C.

6. More or less. The higher one goes in trophic level (from herbivore to predator, from predator to hyper-predator), the higher the relative concentration of phosphorus and nitrogen in the bodies of animals. As a result, a cat that eats a mouse still needs to compensate for the higher concentration of phosphorus in its own body than in that of the mouse. See Angélica L. González, Régis Céréghino, Olivier Dézerald, Vinicius F. Farjalla, Céline Leroy, Barbara A. Richardson, Michael J. Richardson, Gustavo Q. Romero, and Diane S. Srivastava, "Ecological mechanisms and phylogeny shape invertebrate stoichiometry: A test using detritus-based communities across Central and South America," *Functional Ecology* 32, no. 10 (2018): 2448–63.

7. Individual taste receptor cells are also found in the gastrointestinal tract, the sinuses, and even the lungs. These isolated taste receptors help to sort the world's bits into good and bad, but they do so differently. They do not trigger conscious taste sensations. Their role appears to be to help tell the body how to deal with particular compounds rather than whether or not to ingest the food. Paul A. S. Breslin, *Chemical Senses in Feeding, Belonging, and Surviving: Or, Are You Going to Eat That?* (Cambridge University Press, 2019).

8. Contrary to what you probably learned in high school biology, different parts of your tongue do not taste different kinds of things. Each taste bud appears to contain, in and amongst its petal-like cells, each and every type of taste receptor.

9. To make the fish flakes, the muscles of a fish, typically the skipjack tuna (*Katsuwonus pelamis*), also known as bonito or, in Japanese, *katsuo*, are gently boiled in salt water for an hour. The boiled fish muscle, *sans* skin, is next smoked over a hardwood fire for twenty days and twenty nights. The smoked pieces of meat are seeded with fungal spores, often including multiple species of *Aspergillus*, *Eurotium*, and *Penicillium*.

The inoculated fish is put in a sealed box to ferment. After a few days, the resulting mold is scraped off the fermented, smoked meat. The meat is allowed to ferment again. This process of fermentation and scraping is repeated five times during the course of a month. After the fifth scraping, the fish is done. The smoked, fermented fish is then made into flakes called *katsuobushi*. These flakes are the first ingredient in dashi.

10. However, the concentration of simple sugars necessary to trigger sweet taste receptors varies with the body size of an animal species. Smaller animal species have faster metabolic rates and so need a higher concentration of sugar to fuel their body. Only very, very, sugary nectar or fruits are perceived as sweet by the small mammals, such as the smallest monkeys. Big mammals can rely on lower concentration sugars because they need less energy per pound of body mass (and so also less sugar per bite of food) than do smaller mammals. Also, larger animals have longer guts and so can rely more on complex carbohydrates that take their gut and gut microbes longer to break down and convert to energy. As a result, to an elephant, even a leaf of grass can be sweet. As humans, we fall in the middle of this spectrum. Some of what is sweet to us is not sweet to, say, marmosets. Meanwhile, things that evolved to attract smaller mammals with sweetness are wonderful to us.

11. As recently reported in an interview with documentary film maker Annamaria Talas.

12. Recent research has suggested that some fatty acids trigger tastes. Fats and oils are triglycerides, made up of three fatty acids linked to each other by a molecule of glycerol. When fats begin to break down, for example during decomposition, these fatty acids are freed from the glycerol and from each other. Some very short fatty acids trigger the sour taste receptor and so taste sour (acetic acid is a very short fatty acid). Medium-length fatty acids, however, have their own taste. The hard-to-describe taste of these medium-length fatty acids is displeasing. Rick Mattes and colleagues call this taste "oleogustus" (*oleo* meaning oily or fatty in Latin and *gustus* meaning taste). Cordelia A. Running, Bruce A. Craig, and Richard D. Mattes, "Oleogustus: The unique taste of fat," *Chemical Senses* 40, no. 7 (2015): 507–16.

13. This warning system has been used by governments to promote human safety. For example, the most bitter compound yet discovered, denatonium benzoate, is often added to household cleaning supplies and pesticides to warn anyone that accidentally ingests such products that they are dangerous.

14. The stick is stronger in children. Children are more responsive to bitter flavors, such as coffee, chocolate, and beer that contains hops, than are adults. We don't know how this occurs in the brain, how taste is modulated with age, but it is. It may be that this stronger aversion to bitter and potentially toxic foods among young people is adaptive and evolved to help protect young people who are slightly more prone to both encounter new foods and to ingest them without knowing better.

Young people are also attracted to higher concentrations of sugar and salt. In general, the young tongue, in its admonishments, shouts louder. "OVER HERE. NO. NO. NO. NOT OVER THERE." See, for example, J. A. Mennella, M. Y. Pepino, and D. R. Reed, "Genetic and environmental determinants of bitter perception and sweet preferences," *Pediatrics* 115, no. 2 (2005): e216–e222.

15. In such statements, Lucretius made no distinction between humans and other animals.

16. But there is a little more here too. Somehow, despite lacking umami taste receptors, pandas are able to find and choose the species of bamboo and the parts of bamboos that are highest in protein. For most of the year, they eat the leaves of their primary bamboo species. But when the shoots become available, they switch to the shoots which have much more protein. Finally, when the shoots and leaves of their primary bamboo species decline in protein content, the pandas migrate to a high elevation site and eat the shoots of another species of bamboo that, at that time of year, contain more nitrogen. The pandas, in essence, eat these tall grasses at their meatiest. What is not known is how they know they are at their meatiest. One possibility is that one of their taste receptors has evolved to be able to detect amino acids that are common in the bamboo and, hence, indicative of protein. This possibility has not been tested. Yonggang Nie, Fuwen Wei, Wenliang Zhou, Yibo Hu, Alistair M. Senior, Qi Wu, Li Yan, and David Raubenheimer, "Giant pandas are macronutritional carnivores," *Current Biology* 29, no. 10 (2019): 1677–82.

17. Many of the key insights in this chapter were developed through an ongoing, and extremely rewarding, collaboration with Mick Demi, Brad Taylor, and Ben Reading. Michael Tordoff, Stan Harpole, Jon Shik, Matthew Booker, Chad Ludington, Rick Mattes, Carlos Martinez del Rio, Wei Fuwen, Annamaria Talas, Karen Kreeger, Dani Reed, Lee Fratantuono, David Norbrook, and Neil Shubin read the chapter and provided valuable insights. Kim Wejendorp and Josh Evans helped to provide a culinary perspective on taste. We'd also like to thank Mike Kaspari, whose insights into the elemental needs of organisms were key to our thinking about stoichiometry and taste.

Chapter 2. The Flavor-Seekers

1. What is more, four-thousand-year-old rocks and hammers, used by chimpanzees now long gone, were recently found deep in the soil at the same site. Not only do different chimpanzee populations have unique culinary traditions, in some cases those traditions and their tools may be thousands of years old.

2. We can imagine it starting with our last common ancestor with chimpanzees and ending once our ancestors begin to use sharp, stone tools.

3. A chimpanzee spends about four times as much energy knuckle walking from place to place as we spend walking. For an elegant description of the changes that

occurred to our ancestors' bones during these evolutionary transitions, read Daniel Lieberman, *The Story of the Human Body: Evolution, Health, and Disease* (Vintage, 2014).

4. Here we need to define a few more terms because not all roots are really "roots." As the botanist Chris Martine put it to us in an email, "In the most basic sense of the plant body plan, you've got just two things: the shoot system and the root system. Shoot systems are normally above-ground and consist of stems, leaves, and flowers. Roots are normally below-ground and serve to anchor, move stuff in and out of the plant and as a storage system. Many plants store energy in their roots. Every now and again plants will also use some part of the shoot system as another way to do below ground storage. One version is the tuber, really just a section of stem that is swollen and fat with stuff. We know it's a stem because it has buds on it (e.g., the eyes of a potato) as opposed to a storage root that lacks any buds (e.g. the eye-free sweet potato). Another version is the bulb, an aggregation of swollen storage leaves (or at least the bases of the leaves)." As a result, the different things that a chef might call "root vegetables," are really very different parts of the plant. Yet, despite their botanical differences these parts (roots, tubers, and bulbs) play similar culinary roles and so we tend to use the term "root," in the broad culinary way here (botanists would prefer we say "underground storage organ" which is unwieldy and also a little vulgar-sounding), except when we seek to call attention to just tubers or just bulbs. No doubt this will frustrate some botanists. We are sorry. We love botanists.

5. This period of our evolution in which our ancestors and those chimpanzees diverged remains mysterious. Much of it occurred in forests and forest edges where fossils preserve poorly. As Daniel Lieberman notes, all of the fossil hominins from this period would fit in a grocery bag.

6. With names differing depending upon the expert one consults. For example, *Australopithecus robustus* is sometimes referred to as *Paranthropus robustus*. Whatever its name, it was somehow closely related to (and yet distinct from) the other *Australopithecus* species.

7. The need of *Australopithecus* species for patches of trees was highlighted in a recent study by Amanda Henry at the University of Leiden of the diet of two individuals of one species of *Australopithecus sediba*. Those individuals were found in a habitat in which the other species present were big grazing mammals and grassland plants. Yet, the microscopic bits of plants stuck in the teeth of the *Australopithecus* were bits of what appeared to be nuts, shells, leaves, and bark from trees. In addition, the form of carbon in the teeth was the kind that would be expected from eating forest food. Surrounded by grassland, these two individuals lived off the woods. Amanda G. Henry, Peter S. Ungar, Benjamin H. Passey, Matt Sponheimer, Lloyd Rossouw, Marion Bamford, Paul Sandberg, Darryl J. de Ruiter, and Lee Berger, "The diet of *Australopithecus sediba*," *Nature* 487, no. 7405 (2012): 90.

8. A great deal of paleoanthropology is about teeth. Even their subtle differences tell stories, whether with regard to their chemical composition, size, shape, or wear. But we know that not everyone finds ancient teeth super exciting. Take our kids, for example. We were excited, recently, to show them a tooth preserved in a museum outside Guadix, Spain, that appears to be from an ancient human. The tooth is no fewer than a million years old and, some argue, closer to two million years old. It is astonishing, a sublime token of the past, or at least that is what we kept telling our kids as they wandered over to look at something else, anything else.

9. This works, in part, by blocking the olfactory receptors in the antennae of the bees. The bees can't smell the person who is coming to gather the honey. But they also can't smell the alarm pheromone, isopentyl, that the first bees to see, feel, or smell the honey-gatherer release. P. Kirk Visscher, Richard S. Vetter, and Gene E. Robinson, "Alarm pheromone perception in honey bees is decreased by smoke (Hymenoptera: Apidae)," *Journal of Insect Behavior* 8, no. 1 (1995): 11–18.

10. In this way, the calorie labels on raw, unprocessed foods you buy in the store are a kind of lie. Those labels indicate the number of calories you would get from a food if you digested it fully, but the completeness of digestion depends on how you prepare the food as well as the details of the species of microbes in your gut.

11. Wrangham is quick to acknowledge that one of the people to have advocated for a similar idea is Brillat-Savarin, who wrote that "It is by fire that man has tamed nature." Meat, Brillat-Savarin goes on to argue, becomes more desirable and valuable once it is cooked.

12. In addition, as the anthropologist Alyssa Crittenden pointed out to Rob, many of the ways in which modern foragers use fire would never turn up in the archaeological record. As Alyssa put it, "Contemporary foragers, like the Hadza, make ephemeral fires. We simply can't predict whether hearths (as the Hadza use them—three rocks and a flame) would have left any archaeological signature." See, for example, Carolina Mallol, Frank W. Marlowe, Brian M. Wood, and Claire C. Porter, "Earth, wind, and fire: Ethnoarchaeological signals of Hadza fires," *Journal of Archaeological Science* 34, no. 12 (2007): 2035–52.

13. The chimpanzees at Mahale have been shown to consume about a third of the plant species around them, similar to the proportion of plant species eaten by chimpanzees at Gombe. Their most common dietary item is fruit, but they also eat flower blossoms, leaves, insect galls, bark, pith, and resin. Because the culinary traditions of the Mahale chimpanzees are different than those of the Gombe, however, the identity of those consumed species is different. Only sixty percent of the plant species consumed at Gombe are also eaten at Mahale even though the plant species that are present at the two sites are very similar.

14. Chimpanzees ate, for example, the red fruit of the plant *Pycnanthus angolensis*, which likely evolved to attract birds. To Nishida, this fruit was bitter, woody, and

strange, or, as Richard Wrangham put it, impossibly unpleasant. When Nishida tasted it, he immediately spat it out. Fruits like *Pycanthus angolensis* might have been bitter to Nishida, but not to the chimpanzees. It was also possible, however, that the chimpanzees had learned to like some plant parts despite their bitterness, much the way that humans use hops to flavor beer (and come to enjoy the flavor of hops) or drink coffee. The same fruit is also popular in other chimpanzee populations, including the Rió Muni population in Equatorial Guinea, where Jordi Sabater Pi did his research. Sabater Pi, "Feeding behaviour and diet of chimpanzees (*Pan troglodytes troglodytes*) in the Okorobiko Mountains of Rio Muni (West Africa)," *Zeitschrift für Tierpsychologie* 50, no. 3 (1979): 265–81.

15. That isn't to say humans wouldn't eat the fruits chimpanzees eat. At the site at which Nishida worked one of the fruits most often eaten by the chimpanzees was a fruit of the genus *Saba*, called the bungo fruit. This same fruit is often used by humans across Africa to make juices, juices said to have a flavor that is a little like mango, a little like orange, a little like pineapple. Another common food was the fruit of the tree *Pseudospondias microcarpa*. Again, this is a fruit that local people also frequently eat (though Richard Wrangham notes that it is a fruit best consumed in small doses). The chimpanzees at Mahale also ate *Harungana madagascariensis* which is eaten by humans too, as a snack. And, as we've noted, the chimpanzees ate figs, lots of figs, figs of a half dozen species, some of them perceived to be quite tasty by humans. Nor is it just chimpanzees. In Uganda, mountain gorillas and humans both like omwifa fruits so much that when these sweet fruits are in season humans and gorillas travel to the same sites to gather and enjoy the fruits. J. Sabater Pi, "Contribution to the study of alimentation of lowland gorillas in the natural state, in Río Muni, Republic of Equatorial Guinea (West Africa)," *Primates* 18 (1977): 183–204.

16. The story appears to be similar for gorillas. For example, Jordi Sabater Pi, a Catalonian primatologist, observed lowland gorillas in Río Muni in Equatorial Guinea for more than six hundred hours in the 1950s and '60s. While doing so, he, like Nishida, ate what they ate. Sabater Pi discovered that the gorillas preferred fruits that were sweet or sweet and sour, but often had to eat fruits that were insipid because none of their preferred fruits were available. The gorillas of Río Muni do not use tools, but Sabater Pi observed that big, fat, older gorillas, too heavy to climb well, would sometimes urge younger gorillas up into trees to break off branches with desirable fruit and throw them down.

17. Guevara imagines that it was the female gorillas with the mutation that were somewhat better nourished than females without the mutation. The females with the mutation spent more time eating fruits with real sugars instead of wasting foraging time on this fruit. Because the fertility of wild female mammals is tightly linked with energetics, this nutritional edge could result in a fertility edge that could translate over a lifetime into more babies and over generations into fixation of the

mutation in the gorilla population. Elaine E. Guevara, Carrie C. Veilleux, Kristin Salton-stall, Adalgisa Caccone, Nicholas I. Mundy, and Brenda J. Bradley, "Potential arms race in the coevolution of primates and angiosperms: Brazzein sweet proteins and gorilla taste receptors," *American Journal of Physical Anthropology* 161, no. 1 (2016): 181–85.

18. For example, Vittoria Estienne has been studying the honey gathering of chimpanzees at one particular field site, Loango, in Gabon. Here, as in nearly every other chimpanzee population, the chimpanzees use sticks to gather honey out of honey-bee nests. They also use sticks to gather honey out of the nests of stingless bees in trees and a separate species of stingless bees that nests deep underground. When pursuing the latter, Estienne has shown, the chimpanzees will begin to dig into a bee nest and then abandon the digging. The digging takes longer than the chimpanzees are willing to invest in one sitting. Also, as Estienne notes, the chimpanzees get distracted. By other foods. By sounds. By sexy chimpanzees. (In one video from Estienne's field site a male chimpanzee is digging, digging, digging at a bee nest in the ground and then a female chimpanzee with estrus swellings walks by. The male suddenly forgets all about the honey and disappears after her down the trail.) But eventually the chimpanzees come back to the project. Sometimes the same chimpanzee. Sometimes different chimpanzees dig at the same nest. All of this digging can take up to five years and many tens of hours of work to succeed, depending on the hardness of the ground, the depth of the bees, and other factors. And when it is all done and the chimpanzees get to the bee nest, they gather up in their hands the brood and the honey and share them with whoever is nearby. The chimpanzees, after all this work, share and enjoy their spoils, spoils that Estienne has not tasted but that are undoubtedly very sweet (because of the honey), fatty (because of the brood), and even a little bit savory. No one has calculated the precise amount of work it requires to harvest these ground bees, but it is many times more calorically expensive than the reward. The inescapable conclusion seems to be that the chimpanzees continue to work to get the bees because the bees and their honey, together, taste delicious. Vittoria Estienne, Colleen Stephens, and Christophe Boesch, "Extraction of honey from underground bee nests by central African chimpanzees (*Pan troglodytes troglodytes*) in Loango National Park, Gabon: Techniques and individual differences," *American Journal of Primatology* 79, no. 8 (2017): e22672.

19. Personal communication, Maureen McCarthy.

20. Chimpanzees might also make other dietary choices related to flavor, tradition, and social dynamics that do not benefit their fitness. For example, while early studies of predation by chimpanzees on mammals, such as colobus monkeys, emphasized the nutritional benefits of such hunting, more recent studies are more equivocal. A recent paper by Claudio Tennie and colleagues at the University of Birmingham, after much hemming, hawing, and calculating, could not find a clear nutritional advantage to the chimpanzees for eating meat. This isn't to say that a

benefit doesn't exist (the authors thought that more data and analysis might reveal one), but instead that it is far less clear-cut than one might imagine, and likely to vary among sites and scenarios such that hunting and eating is sometimes a complete waste of energy. Claudio Tennie, Robert C. O'Malley, and Ian C. Gilby, "Why do chimpanzees hunt? Considering the benefits and costs of acquiring and consuming vertebrate versus invertebrate prey," *Journal of Human Evolution* 71 (2014): 38–45.

21. Daniel Lieberman noted in an email that the *Crematogaster* ants are especially good, "quite tasty."

22. Perhaps a little more surprising, the Hadza ranked the five main berries they consume (and that men and women gather together) as identical to each other in terms of the quality of their flavor. For the Hadza, one berry is substitutable for another. Tubers, meanwhile, are ranked below berries, but do not rise to the category of distasteful, a category that includes a variety of foods that Hadza do not tend to like, foods they describe as "tasting like snake." Similarly, according to Brillat-Savarin, the French call some disliked animal foods "bêtes puantes." Included on the French list of such stinking beasts are, at least according to Brillat-Savarin, foxes, crows, magpies, and wildcats.

23. One of the few dishes that contains raw mammal meat is steak tartare, but this delicacy requires meat that is chosen specifically so as to contain little connective tissue (a luxury our ancestors did not have) and then also minced (further improving its mouthfeel). In addition, it is accompanied by eggs, onions, and sauces for flavor.

24. Of course, in each thing moderation. Brillat-Savarin noted that "[people] who eat quickly and without thought do not perceive the [succession of] taste impressions, which are the exclusive perquisite of a small number of the chosen few; and it is by means of these impressions that gastronomers can classify, in the order of their excellence, the various substances submitted to their approval." Brillat-Savarin was right. Recent studies have shown that to experience the full flavors of food one must chew slowly. But "slowly" is relative. The magic rate is slightly more slowly than most folks do and yet much more briefly than does a chimpanzee.

25. Here we pause for a caveat. Raw meats are not enticing to most modern humans. We infer that they also weren't very enticing for our ancestors. However, two different researchers whose work focuses on chimpanzees in the field, Hjalmar Kuehl and Mimi Arandjelovic, noted that when chimpanzees have killed a monkey, the enthusiasm they show for tearing it up and eating it looks like pleasure. It looks just like pleasure. Hjalmar and Mimi both, independently, posited that perhaps the chimpanzees have some sense of taste or mouthfeel that we lack. We can't rule out such a possibility. However, what we do know is that even if chimpanzees like raw meat more than humans do, they like cooked meat more than they like raw meat.

26. Our understanding of the relationships between the recent species of *Homo* that lived during the last million or so years is becoming ever clearer, thanks in large

part to the recovery of ancient DNA and protein from fossil teeth and bones. However, the relationships among the different species of *Homo* that lived between 1.9 million and 800,000 years ago remains contentious. If you want to read more about the most recent studies of ancient proteins of hominins and what they say about the emerging relationships among them, see a beautiful recent paper in which Frido Welker and colleagues were able to extract and study protein from teeth of a roughly 800,000-year-old species of *Homo* from Spain (that they call *Homo antecessor* and we lump into *H. erectus*). That ancient human was roughly as different from modern humans as modern chimpanzees (*Pan troglodytes*) are from modern bonobos (*Pan paniscus*), which is to say different and yet not so very different. Frido Welker, Jazmín Ramos-Madrigal, Petra Gutenbrunner, Meaghan Mackie, Shivani Tiwary, Rosa Rakownikow Jersie-Christensen, Cristina Chiva, et al., "The dental proteome of *Homo antecessor*," *Nature* 580, no 7802 (2020): 1–4.

27. This is not to say no differences in receptors exist among *Homo sapiens* or among human species, but instead to suggest that such differences are modest compared to the similarities. In addition, some differences found in taste receptors found within *Homo sapiens* are also found within other human species. For example, it has long been known that humans vary with regard to whether they are tasters or nontasters of phenylthiocarbamide. To some people this compound tastes bitter. To others, it has no taste at all. But this variation does not just exist within humans. A recent study has shown that some Neanderthals were tasters and some were nontasters for this compound. In other words, with regard to taste receptors even our differences are ancient and shared with other species of humans. Carles Lalueza-Fox, Elena Gigli, Marco de la Rasilla, Javier Fortea, and Antonio Rosas, "Bitter taste perception in Neanderthals through the analysis of the TAS2R38 gene," *Biology Letters* 5, no. 6 (2009): 809–11.

28. Daniel Lieberman, Alyssa Crittenden, Colette Berbesque, David Tarpy, Becky Irwin, Thomas Kraft, Aung Si, Hjalmar Kuehl, Vittoria Estienne, Christophe Boesch, Katie Amato, Matthew Booker, Chad Ludington, Ran Barkai, Jack Lester, Maureen McCarthy, Carles Lalueza Fox, Mimi Arandjelovic, Amanda Henry, Roman Wittig, Ammie Kalan, Michael Tordoff, Matthew McLennan, Joanna Lambert, and Charlie Nunn all read, commented on, and/or talked about this chapter at length. Richard Wrangham helped clarify ideas. Kim Wejendorp, Josh Evans, Ole Mouritsen, and Michael Bom Frøst all read the chapter and helped add a culinary perspective.

Chapter 3. A Nose for Flavor

1. Black truffles (*Tuber melanosporum*) are typically found in the Dordogne region of southwestern France. White truffles (*Tuber magnatum),* on the other hand, are found in northern and central Italy.

2. Zebra fish have a special kind of olfactory receptor that recognizes cadaverine and another for putrescine. Once stimulated, these receptors trigger an innate aversive reaction in the fish. The senior author of one study, Sigrun Korsching, thinks it plausible that humans also have such a receptor—plausible but untested. Ashiq Hussain, Luis R. Saraiva, David M. Ferrero, Gaurav Ahuja, Venkatesh S. Krishna, Stephen D. Liberles, and Sigrun I. Korsching, "High-affinity olfactory receptor for the death-associated odor cadaverine," *Proceedings of the National Academy of Sciences* 110, no. 48 (2013): 19579–84.

3. Interestingly, many butterflies and moths use this same compound, mixed with other compounds, for their own attractant. This similarity leads to two observations. One is that some compounds appear to be more effective as pheromones than are others, whether because they travel greater distances, stay longer in the environment, or are easier for noses and antennae to detect. The other is that male Asian elephants may find some moths to be sexy (though because of issues of quantity, the reverse is probably even more true, that some moths are drawn to male elephants like, well, moths to male elephants). David R. Kelly, "When is a butterfly like an elephant?" *Chemistry and Biology* 3, no. 8 (1996): 595–602.

4. In most cases it is difficult to predict how a chemical will smell based solely on its structure. Not so for compounds that contain disulfide bonds. Disulfide bonds occur when two molecules (often two proteins) are connected to each other by a bond between two sulfur atoms. Such compounds invariably smell like garlic, rotten cabbage, or must. Andreas Keller and Leslie B. Vosshall, "Olfactory perception of chemically diverse molecules," *BMC Neuroscience* 17, no. 1 (2016): 55.

5. In this way, the relationship between dogs and the kitchen is ancient, but ever changing. At one time dogs helped to hunt for our most delicious and cherished foods, whether truffles or mastodons. Now they eat our scraps, whether the scraps from cooking, the scraps from the table, or the bits and pieces of fish and other animals we don't ourselves enjoy very much, ground up into a kind of food mulch and canned.

6. As a chef told us in conversation, "Anyone can win a Nobel, only one person lent their name to the deliciousness of cooked food."

7. The browning happens more rapidly if the pH of the food being cooked is increased (made more alkaline). This is why German pretzels (laugenbretzeln) are treated with lye before cooking.

8. A similar reaction occurs when milk is cooked. At high temperatures, for instance, the lactose in milk interacts with proteins and yields butterscotch flavors. These flavors can be featured if milk is brushed onto pastries before they are cooked.

9. As Hsiang-ju Lin and Tsuifeng Lin put it, "The curious, omnivorous cook knows that the taste of raw fruit is quite delicious and cannot be improved upon."

10. It is often suggested that humans have a "degenerate" sense of smell. By some measures, this is true. We have fewer kinds of olfactory receptors and fewer

individual receptors than do, say, our proto-primate ancestors and many fewer than do dogs. In general, as our eyes and brains became bigger, we tended to lose kinds and quantities of those receptors (humans have fewer than chimpanzees, which have fewer than monkeys, which have fewer than lemurs). However, we have much more brain dedicated to making sense of what the olfactory receptors detect.

11. The first such smell map produced was for cheddar cheese. Gordon Shepherd, author of *Neurogastronomy*, went to the store and bought a hunk of cheddar cheese. He then fed it to rats and then, after killing the rats, looked at their brains. In doing so, he saw, for the first time, the "cheddar" constellation.

12. It is not just us speaking in metaphor here. The combinations of receptors triggered by an individual compound are indeed called, by the scientists who study them, "olfactory receptor codes." Ji Hyun Bak, Seogjoo Jang, and Changbong Hyeon, "Modular structure of human olfactory receptor codes reflects the bases of odor perception," *BioRxiv* (2019): 525287.

13. The same is true for the adjectives used to describe the aromas and flavors of foods. They are individual. For example, one study compared the words used by four of the world's best wine tasters to describe wines. First of all, they used words that didn't really even describe aromas, for example, to say a wine's aroma was "red" or "great" or "honest." But in addition, the four individuals, although they were very consistent in the terms they would each use for a particular wine (they could taste it again and again and describe it similarly), used totally different terms from each other. Collectively, they used four thousand different words to describe wine, and of these, the only words they all used were "black currant" and "dark." Shepherd, *Neuroenology: How the Brain Creates the Taste of Wine* (Columbia University Press, 2016).

14. One exception, albeit a modest one, seems to be that most languages studied to date include as aroma categories terms that denote "sweat and body odor," "strong animal smell [which is to say, the sweat and body odor of other species]" and "rotten things odor." C. Boisson, "La dénomination des odeurs: Variations et régularités linguistiques," *Intellectica* 24, no. 1 (1997): 29–49.

15. Farther south, at Gorham's Cave in Gibraltar, the hidden flavors that were being revealed were more varied: charred acorns, pistachios, peas, and legumes, along with ibex, rabbit, red deer, limpets, cockles, mussels, tortoise, monk seal, dolphin, and pigeon. Kimberly Brown, Darren A. Fa, Geraldine Finlayson, and Clive Finlayson, "Small game and marine resource exploitation by Neanderthals: The evidence from Gibraltar," in *Trekking the Shore* (Springer, 2011), 247–72.

16. This chapter was read and helped along by Daniel Lieberman, Gordon Shepherd, Sylvie Issanchou, Benoist Schaal, Mimi Arandjelovic, Sigrun Korsching, Natasha Olby, Roland Kays, Mary Jane Epps, Ran Barkai, Susann Jänig and John Meitzen. Once again, Josh Evans and Kim Wejendorp added a culinary perspective, as did Harold McGee.

Chapter 4. Culinary Extinction

1. Harrison died at his table in Patagonia, holding his pen. He lived on the land, walking it, writing about it, and eating of it. Harrison studied, hunted, and savored many of the animal species that could be found around his Patagonia home. We would have loved to have walked the hills around Patagonia with Harrison, talking about the flavors of wild mammals. But we were too late. Harrison was already dead. We were also too late for the feast thrown after Harrison's death in honor of his life and excursions. Seventy-two people attended, including some of Harrison's closest friends. Nearly as many ducks were present at the table. The menu featured a duck pâté appetizer. This was followed by a cassoulet that took eight days to prepare and contained duck meat, pork sausage, and white beans, all topped with the special duck fat found between the duck's muscle and its skin. For those who still had room, a bisque of grouper, snapper, and shrimp in a tomato broth flavored with fennel, saffron, and Pernod was available, as was a Waldorf salad meant to evoke Harrison's Michigan roots, and French wine. All of this was followed by cakes, more wine, and cigarettes. https://www.outsideonline.com/2291316/behind-scenes-jim-harrisons-farewell -dinner.

2. They were also accompanied by a beautiful and well-crafted "wrench" made out of a mammal leg bone. No good explanation for its use has yet been proposed.

3. That even simple roasting takes learning is highlighted by the myth of how the Wyandot people found fire: "The creator caused fire to gush forth and directed the first man to put a portion of the meat on a stick and to roast it before the fire. But the man was so ignorant that he let it stand until it was burnt on one side, while it was raw on the other." Claude Lévi-Strauss, "The roast and the boiled" (1977), in J. Kuper, ed., *The Anthropologist's Cookbook* (Routledge, 1997), 221–30.

4. The various peoples descended from the Clovis would come to prefer very different styles of cooking meat. As Levi-Strauss notes in *The Raw and the Cooked*, the Assiniboine preferred roast meat to boiled meat, but liked both rare. The Blackfoot, on the other hand, roasted meat, then quickly blanched it in hot water. The Kansa and Osage, meanwhile, preferred their meat very, very well done. The Cavineño in Bolivia boiled food overnight, sometimes even putting meat from the next day in with that of the last to keep the dish going (an approach with kinship to the French cassoulet). Claude Lévi-Strauss, *The Raw and the Cooked* (University of Chicago Press, 1983).

5. So big and terrifying that they are described by some paleontologists, according to the author Craig Childs, as "holy mother of god bison." Childs, *Atlas of a Lost World: Travels in Ice Age America* (Vintage, 2019).

6. It is not the only menu that speaks to what is lost. The plant silphium, for instance, was a favorite of the Romans. It appears to be gone. Today, you cannot taste,

as Adam Gopnik notes, the "delicacy of sea squirt and silphium." Silphium grew in Cyrene, the north of what is now Libya. It was highly desired as a spice. It appears that silphium was harvested into extinction. It tasted, we are told by humanists, a little like asafoetida (which, in turn, tastes a little like rotting garlic). Balsam of Mecca also appears to be gone. So too tejpat.

7. The area around the Clovis sites in southern Arizona, for instance, changed (as Paul Martin himself had shown) from grassland to forest. That change was undoubtedly difficult for grassland and tundra-loving species, such as mammoths, but might also have been expected to favor other species, such as mastodons, which preferred forests and tree fruits and leaves.

8. For example, one recent study concluded that cold-loving woolly mammoths (just one mammoth species from among a small menagerie of forms) were greatly affected by warming conditions, and that their range became restricted to the coldest climates of North America. But, then, once their populations were reduced by warming, hunting likely had a much greater effect than it would have otherwise had. D. Nogués-Bravo, J. Rodríguez, J. Hortal, P. Batra, and M. B. Araújo, "Climate change, humans, and the extinction of the woolly mammoth," *PLoS Biology* 6, no. 4 (2008).

9. It seems reasonable to hypothesize that the same is true of all animals. To our knowledge, however, no one has considered the ways in which the decisions of nonhuman animals about what to eat are influenced by the flavors of their options.

10. The people Koster interviewed noted another reason that they did not eat carnivorous species besides their flavor. Carnivores, be they jaguars (*Panthera once*) or tayras (*Eira barbara*), eat raw meat. Koster notes that taboos against eating animals that eat meat are common. Recently, it has even been suggested that carrion-feeding mammals tend to avoid the meat of carnivores. One explanation of why this is is that such carnivores are more likely to have parasitic worms and other pathogens (from their prey) that might cause harm to those who eat them. Marcos Moleón, Carlos Martínez-Carrasco, Oliver C. Muellerklein, Wayne M. Getz, Carlos Muñoz-Lozano, and José A. Sánchez-Zapata, "Carnivore carcasses are avoided by carnivores," *Journal of Animal Ecology* 86, no. 5 (2017): 1179–91.

11. The exception being pacas, which appear to reproduce more quickly than they are eaten. Jeremy Koster, "The impact of hunting with dogs on wildlife harvests in the Bosawas Reserve, Nicaragua," *Environmental Conservation* 35, no. 3 (2008): 211–20.

12. Though it can take on more flavors when cooked with bone.

13. One of the exceptions that makes the point is seeds. Plant seeds need to be small and mobile and tend to store their energy in fat, a reality we take advantage of every time we use canola oil or sesame oil.

14. Koster reports that monkey muscles become fattier during the rainy season and that hunters throughout the Americas look forward to "rainy-season meat," but

that during the dry season that monkey meat is leaner, which appears to affect flavor. For example, the Piro and Machiguenga in Peru contend that lean, dry-season primates are not worth pursuing.

15. In addition, some vertebrate species have special kinds of fats. For example, as the ornithologist Jon Fjeldså noted in an email, "Some seabird meat tastes of fish oil." But, as he and his colleagues figured out, "this taste could be removed by removing all fat as soon as possible after they were shot. This was essential with cormorants, where the subcutaneous fat has a very low melting point and runs off like oil as soon as you pull off the skin. . . . [T]his has to be done within a minute after the bird is shot and then even the cormorants taste quite good." Fjeldså was shooting the birds as part of efforts to document their distribution and biology. After he killed the birds and removed the parts of their bodies needed for his studies, he was always eager to use the meat so that nothing would go to waste.

16. In addition, many predators have musk glands that, if they are not carefully removed, lend a musky flavor to the meat that no one seems to like.

17. This may be especially true for omnivores, such as many bear species, whose ancestors were predators. Such species still have quite simple guts, predator guts. As a result, their meat can readily pick up the flavors of what they have eaten. This appears to be the case for grizzly bears. In an email Gary Haynes described grizzly bear meat as "tasting like roots and rodents, which is to say, not delicious. Yuck." Conversely, the paleoarchaeologist Todd Surovell, in reading this note about Gary's experience, described a very different meal he consumed in Mongolia in which he was given and ate a delicious piece of bear meat (from a subspecies called the "black grizzly") which his companions noted, "tasted of berries and pine nuts."

18. Via email, February 28, 2020. Fjeldså found a pine grosbeak (*Pinicola enucleator*) to have similarly excellent spicing. The pine grosbeak had flown into a window and died (after being scared by bird watchers). The Danish Natural History Museum took the dead bird for its collection. Its tissues were taken for DNA research, its skin for study. And its meat was, well, eaten. Six people shared "the breast muscles of this 50-gram bird, with a gravy of chanterelles and port wine." All involved "agreed that this was absolutely delicious, and the probable reason for this must be" the bird's diet "of various berries and buds of spicy plants."

19. As well as to the anthropologists working with the Hadza. Colette Berbesque, the first person to study Hadza food preferences, noted that warthog, when cooked by the Hadza, tastes like really delicious ham (email, May 16, 2019).

20. Based on the experience of Jon Fjeldså.

21. The one exception for pigs is that the meat of male pigs, of diverse sorts, can sometimes have a "boar odor." This odor comes from androstenone, one of the two chemicals that make up boar pheromone, the one that smells a bit "urinous." Michael J. Lavelle, Nathan P. Snow, Justin W. Fischer, Joe M. Halseth, Eric H. VanNatta,

and Kurt C. VerCauteren, "Attractants for wild pigs: Current use, availability, needs, and future potential," *European Journal of Wildlife Research* 63, no. 6 (2017): 86.

22. Josh Evans, a former food innovator at Nordic Food Labs and now a geographer focusing on food, notes, in reading this chapter, that things may be very different in insects. In many cultures herbivorous insects are common food items. But, Josh argues, it seems as though the insects that are preferred are those that feed on a relatively narrow range of plants and, in doing so, concentrate the unusual flavors of those plants, as is the case with palm weevils, cherry caterpillars, and tobacco crickets. For more on insects as food, see the beautiful book *On Eating Insects: Essays, Stories and Recipes*, by Joshua David Evans, Roberto Flore, and Michael Bom Frøst (Phaidon, 2017).

23. Early in the discussions of megafauna extinction, the University of Alaska zoologist Dale Guthrie argued that non-ruminants were more likely to go extinct during the extinction than ruminants. He doesn't appear, however, to have considered the possibility that the susceptibility of non-ruminants could have been due to their deliciousness. R. D. Guthrie, "Mosaics, allelochemics, and nutrients: An ecological theory of late Pleistocene megafaunal extinctions," in *Quaternary Extinctions*, ed. P. S. Martin and R. G. Klein (University of Arizona Press, 1984), 289–98.

24. Though we can't know for sure. It is also possible that some mammals have, because of details of their biology beyond those we've considered here, bad flavors. As mammologist Roland Kays pointed out to us when reading this chapter, the extant species of sloth tend to be regarded as having rather terrible flavors. This could be because of their diets (they eat leaves and leaves alone), in which case a giant sloth with a different diet might nonetheless have a lovely flavor. Or, there could just be a "sloth flavor," that big or small is never something one really wants to indulge in if one has a choice. The flavors of the past remain mysterious.

25. Though the populations of these species would have been at risk anyway because of the loss of their prey and, if the Miskito are any indication, because species viewed to be dangerous are sometimes killed even if they are not eaten.

26. Once it is, the muscle can do nothing but wait before firing again, both until the oxygen concentration is restored and until the the lactic acid (a waste product of metabolism) has been removed.

27. However, ecological absolutes are often honored by their exceptions. One chef, in reading this paragraph, wondered if there might have been some proboscidean that specialized on a particular, delicious fruit and, in doing so, had meat with a wondrously unique flavor. This seems unlikely, but far be it for us to ruin a chef's culinary dreams.

28. Though, as with much in the past, there is a caveat. Gary Haynes, in reading this text, noted that there were drought years during much of the Clovis period. Such droughts might have meant that many mammoths began to go hungry. If this were

the case, their meat, Haynes speculates, would have been stringy during such periods. Paleoanthropologist Ran Barkai, on the other hand, pointed out that even a skinny mammoth was probably pretty fat.

29. The Wata killed elephants with long bows, the arrows of which were dipped in poison. Between 63 BCE and 24 CE Strabo documented a similar practice along the Red Sea where elephants no longer roam. He noted that it takes three men to shoot: two to hold the enormous bow while the third pulls the string. This is how the Wata still killed elephants in the early 1900s. The Wata arrows were tipped in poison made from a mix of plants, including trees of the genus *Acokanthera*. Ian Parker, "Bows, arrows, poison and elephants," *Kenya Past and Present* 44 (2017): 31–42.

30. As reported in Reshef and Barkai, "A taste of elephant."

31. It was also in the Dordogne where Neanderthals and *Homo sapiens* overlapped the longest (some six thousand years). During this time, humans and Neanderthals exchanged genes, art, and, we suppose, recipes.

32. For more on these abstractions and related symbols, see Genevieve von Petzinger, *The First Signs: Unlocking the Mysteries of the World's Oldest Symbols* (Simon and Schuster, 2017).

33. Flavor is no longer the best predictor of which mammals remain common in North America. Once the tasty mammals were gone, hunters appear to have picked again from what was left, choosing the tastiest of the leftovers until all that remained were the smallest, fastest-reproducing, least tasty species, a reality now true for much of the world. Rodolfo Dirzo, Hillary S. Young, Mauro Galetti, Gerardo Ceballos, Nick J. B. Isaac, and Ben Collen, "Defaunation in the Anthropocene," *Science* 345, no. 6195 (2014): 401–6.

34. This chapter was read and improved by Harry Greene, Carlos Martinez del Rio, Gary Graves, Jon Fjeldså, Roland Kays, Joanna Lambert, Alston Thoms, Nate Sanders, Todd Surovell, Gary Nabhan, Genevieve von Petzinger, Jeremy Koster, Scott Mills, John Speth, Nate Sanders, Ran Barkai, and Colette Berbesque. It benefited from discussions with Dan Fisher. Once again, Josh Evans and Kim Wejendorp added a culinary perspective.

Chapter 5. Forbidden Fruits

1. The French language is particularly full of fruitful phrases. A compromise is "couper la poire en deux," to cut the pear in two. To pick someone's brain is to squeeze their lemon, "presser le citron."

2. Though with fruits there are always exceptions. One of those exceptions is a tropical understory plant that is dispersed by monkeys. The plant's fruits are enticing, but its seed are bitter and toxic and so monkeys, upon eating the fruits, encounter

the toxic seeds and spit them out. These seeds travel one expectoration at a time. Ian Kiepiel and Steven D. Johnson, "Spit it out: Monkeys disperse the unorthodox and toxic seeds of *Clivia miniata* (Amaryllidaceae)," *Biotropica* 51, no. 5 (2019): 619–25.

3. Leave it to ecologists to look at the different ways in which inanimate fruits woo animals to them from miles away, a near-magical ability of a sessile organism, something so intriguing as to feature in one of the central stories of Western religions, and turn this into something that sounds like a disease.

4. Along with Noah Fierer, Valerie McKenzie, and their daughter, and Anne Madden and Tobin Hammer. We'd actually come in search of a bee reported to make beer in little bowls in which its eggs then hatched, beer for its babies. We didn't find the bee, but we did find Janzen.

5. Horses, we know thanks to feeding trials, prefer sweet foods to those that are not sweet. But they do not prefer foods that are sour or salty. If foods are very salty or very sour, horses reject them. R. P. Randall, W. A. Schurg, and D. C. Church, "Response of horses to sweet, salty, sour and bitter solutions," *Journal of Animal Science* 47, no. 1 (July 1978): 51–55.

6. Janzen wasn't done. He also decided to observe a kind of natural experiment. In parts of the Chihuahua desert of Mexico (west and stretching south, as the Chihuahua raven flies, from Patagonia, Arizona), many kinds of opuntia cactus are native. Opuntia cactuses have large fruits that one can now buy in Mexican food stores in the United States and sometimes in bigger supermarkets (the paddle-like leaves, meanwhile, are sold as "nopales" or "nopalitos"). The fruits look like brightly colored and slightly deformed tennis balls. The opuntia fruits seem likely to have been megafauna fruits. In places in Africa where the opuntia cactuses have been introduced alongside elephants, the elephants prefer their fruits (and spread their seeds). In Chihuahua, Janzen had the idea that it was possible that the opuntia fruits, in the absence of their native megafauna, might be getting dispersed by the roaming cows of Mexican ranchers. But rather than release cows into places where they weren't already roaming (after feeding them opuntia), Janzen did something simpler. He compared the abundance and diversity of opuntia cactuses in areas where cows had been fenced out to where they hadn't. Where cows were fenced out, or simply not allowed to roam, the opuntia cactuses were largely gone. They weren't being dispersed. What was more, what grew in their place were other species, spineless species that the cows (modern replicas of ancient desert herbivores) would have otherwise eaten. Janzen's observations of opuntia cactuses were more evidence both for his megafauna disperser idea and for the idea that with the megafauna of the Americas gone, other species can, in certain circumstances, take on some of their roles.

7. In other cases, survival might reflect very specific features of the biology of a species. The pawpaw is a tree with a delicious megafauna fruit that grows across

much of eastern North America. Related to the custard apple, its fruits taste like a mix between a banana and a mango. But their seeds tend to go undispersed in forests. One way the tree has persisted is by water dispersal. It now grows along rivers even though such riparian habitats aren't ideally suited to the tree's growth. But more recently, it has become more successful. The leaves of the tree are distasteful to deer. Where deer are common, the pawpaw now sometimes grows in dense stands, dense, delicious stands.

8. This chapter was greatly improved by the reading and insights of Maarten van Zonneveld, Doug Levey, Omar Nevo, Renske Onstein, Elaine Guevara, Gregory Andersen, Christopher Martine, Gary Haynes, Joanna Lambert, Robert Warren, Lisa Mills, and Thomas Kraft. Once again, Josh Evans and Kim Wejendorp added a culinary perspective.

Chapter 6. On the Origin of Spices

1. The only possible exception is that chimpanzees sometimes chew leaves while eating meat. This might be a kind of spice use, but chimpanzees use whatever plants are at hand as additions to the "bowls" of their mouths. They haven't learned to use a particular spice to make a specific mix in their mouths.

2. Another factor that influences the chemistry of some herbs, such as thyme, is temperature. The defensive chemicals in thyme leaves are, like those of mints, kept in little balls on their leaves. Where it is cold, those little balls sometimes freeze. When they do, their contents leak out onto the thyme plants. Some of the chemical defenses of thyme, specifically the chemicals carvacrol and thymol, are sufficiently powerful that, upon leaking onto the plant, they often kill it. As a result, thyme plants growing in cold regions often produce fewer powerful, aromatic, defensive compounds. J. Thompson, A. Charpentier, G. Bouguet, F. Charmasson, S. Roset, B. Buatois, . . . and P. H. Gouyon, "Evolution of a genetic polymorphism with climate change in a Mediterranean landscape," *Proceedings of the National Academy of Sciences* 110(8) (2013): 2893–97.

3. *Acinos suaveolens*; it is neither a thyme nor a basil and, confusingly, is distinct from another plant called "basil thyme."

4. This study was done the way you might fear it would be done. The mother sheep were fed garlic. Samples were then taken of their amniotic fluid. A "panel of judges" was then asked to take whiffs of fetal blood, maternal blood, and amniotic fluid. Each of these bodily substances smelled of garlic. Dale L. Nolte, Frederick D. Provenza, Robert Callan, and Kip E. Panter, "Garlic in the ovine fetal environment," *Physiology and Behavior* 52, no. 6 (1992): 1091–93.

5. A newborn baby cannot respond to questions, but, like a newborn rat, can express pleasure and displeasure. In 1974, Israeli doctor Jacob Steiner discovered that

the reactions of babies to different tastes could be discerned based on their facial responses. Sour tastes make babies pucker, for instance. Bitterness triggers an open mouth accompanied by attempts to spit the offending food item out. And sweetness produces what Steiner described as a look of relaxation along with an "eager licking of the upper lip" and an ever-so-slight smile, as do umami flavors. Steiner's results have been repeated dozens of times, and sour, bitter, and sweet faces have now come to be used as general measures of the taste preferences of newborn babies. Similar responses are used to judge the preferences of babies for flavors. It was such responses that were compared in the study of the Alsatian mothers.

6. A related maternal effect can even transcend more than one generation. Recently, one study found that mice whose grandparents learned to associate particular aromas with fear are afraid of those same aromas. The grandbaby mice categorize such aromas (whatever they might be) with negative emotions (they ran in a way that bespoke blind panic). Brian G. Dias and Kerry J. Ressler, "Parental olfactory experience influences behavior and neural structure in subsequent generations," *Nature Neuroscience* 17, no. 1 (2014): 89.

7. Bitter leaf is also among the species of plants used by chimpanzees as medicines. In self-medicating, chimpanzees choose the bitter, hairy, strongly aromatic leaves of a small number of tree species (half a dozen out of the hundreds available to them). They tend to do so in the wet season when parasitic worms are more of a problem. They fold up the leaves into a kind of origami pill. The pill is swallowed without chewing. So consumed, these medicinal plant leaves have been shown to help kill some of the worms in the chimpanzees' guts. Chimpanzees have learned to apply such medical treatments in populations across Africa, apparently independently. So too some gorillas. It is reasonable to assume that our last common ancestor with chimpanzees also used plants as medicine (and maybe even bitter leaf in particular). Michael A. Huffman, Shunji Gotoh, Linda A. Turner, Miya Hamai, and Kozo Yoshida, "Seasonal trends in intestinal nematode infection and medicinal plant use among chimpanzees in the Mahale Mountains, Tanzania," *Primates* 38, no. 2 (1997): 111–25; Michael A. Huffman and R. W. Wrangham, "Diversity of medicinal plant use by chimpanzees in the wild," in *Chimpanzee Cultures,* ed. R. W. Wrangham, W. C. McGrew, Frans B.M. DeWaal, and P. G. Heltne (Harvard University Press, 1994), 129–48.

8. The aroma and hence flavor of any particular allium depends upon how much of the alliin is converted to allicin, and also how much time there is for allicin and other compounds to react (and produce additional compounds). Consider garlic, for example. If you chop garlic finely but do not mash it, some but not all of the garlic cells are smashed and only a little allicin is produced. Chopped garlic is relatively mild. If you crush garlic, more of the cells are burst, more allicin is produced. If you puree garlic, the largest quantity of allicin is produced. And if you cook garlic,

without chopping, crushing, grinding, or pureeing it, the enzyme alliinase is partially deactivated. The garlic will be mild, with a hint of allicin, but only just a hint, a reminder of what this small bulb can do.

9. Included in this collaboration were students from Broughton High School and Wake Stem Academy in Raleigh, North Carolina, Ben Chapman, Natalie Seymour, and Tate Paulette. Ben and Natalie are experts on modern food safety. Tate is an archaeologist specializing in ancient Mesopotamia.

10. The tablets came to Yale in 1933 as part of an acquisition that was relatively ambiguous with regard to its provenance. Based on their style of writing, the tablets appear to date to roughly 1600 BCE. They are thought to come from southern Babylonia.

11. Dishes in other regions at the same time were similarly rich in spices, but did not necessarily employ the same spices. For example, researchers have recently studied the plant material present in cooking sites in an archaeological site called Harrapan, associated with the Indus River Valley civilization. The site, which is approximately 4000 to 4500 years old, contains residues of plants that suggest very complex recipes were already being used. The plants included eggplant, turmeric, ginger, mustard seeds, and mango powder, the makings of a very good curry. Andrew Lawler, "The ingredients for a 4000-year-old proto-curry," *Science* 337, no. 6092 (2012): 288.

12. One interesting difference between the spicy compound in chili peppers and that in cinnamon is that the compound in cinnamon, *trans*-cinnamaldehyde, is light (volatile) and drifts into the nose, where it also accounts for the aroma of cinnamon.

13. One might predict, given the relationship between chilies and birds, that black pepper plants are also dispersed by birds. The truth is we don't know. No one has ever studied the seed dispersal of black pepper plants in India, where they are native. It seems very plausible that black pepper seeds are dispersed by birds. Just as with capsaicin, birds don't get hot mouths from piperine. But there is another possibility. The northern tree shrew lives in tropical forests from China all the way northwest to Bangladesh. Tree shrews are not really shrews but instead a sort of simpleminded relative of primates with a fondness for eating both insects and fruits. A research group in China recently discovered they could feed chili peppers to tree shrews. Why they tried this in the first place is not clear. Just like anyone else, scientists get bored. The chili peppers are not native to the range of the tree shrew. The tree shrew never encounters chili peppers in the wild. It was a ridiculous study, but an interesting result. Once the team discovered that the tree shrews ate chili peppers, they studied the TRPV1 gene in the tree shrew. It was broken. The tree shrew cannot taste capsaicin, nor can it taste piperine. The authors went on to argue that perhaps the tree shrew evolved a broken TRPV1 because individuals with the broken version of the gene were able to eat a species of pepper (with piperine) native to where the

northern tree shrew lives. What the authors did not study (or even comment upon) is the broader possibility that all tree shrew species lack the TRPVI gene and that wild pepper plants (species of the genus *Piper*) throughout tropical Asia (where tree shrews are native) are partially or even largely dispersed by tree shrews. Yalan Han, Bowen Li, Ting-Ting Yin, Cheng Xu, Rose Ombati, Lei Luo, Yujie Xia, et al., "Molecular mechanism of the tree shrew's insensitivity to spiciness," *PLoS Biology* 16, no. 7 (2018): e2004921.

14. Which is to say, that pepper-eating in non-human mammals may be confined to homes, farms, and zoos.

15. The ideas in this chapter were developed in collaboration with Tate Paulette, Pia Sörensen, Ben Chapman, Natalie Seymour, and Lauren Nichols, as well as Swarnatara Stremic, April Johnson, and their students. Shinya Shoda, Sylvie Issanchou, Patience Epps, Gary Nabhan, Joanna Lambert, John Speth, Benoist Schaal, Dietland Muller-Schwarze, Susan Whitehead, Shilong Yang, Oliver Craig, Amaia Arranz Otaegai, Kate Grossman, Tate Paulette, Paul Rozin, Dirk Hermans, Harry Daniels, Doug Levey, Yan Linhart, Rob Raguso, and Ben Reading all read versions of the chapter and helped it along. Once again, Josh Evans and Kim Wejendorp added a culinary perspective.

Chapter 7. Cheesy Horse and Sour Beer

1. Scientists have, however, begun to understand how the sour taste receptor works. In a major breakthrough, Emily Liman, Yu-Hsiang Tu, and colleagues have recently discovered that the sour taste receptor is a protein called OTOP1. They have also discovered that this protein forms a proton channel, a kind of small door that only allows protons through. This door appears to know a food is acidic (and registers it as "sour") if many protons go through the door. Yet, our understanding of sour taste receptors remains partial. It is not known how and why the channel responds differently to organic and inorganic acids. And it is not understood what the sour receptor is doing in other places in the body like the ear or fat tissue. No one knows, yet. But if we had to bet on who will figure it all out it will be Liman and Tu. Tu, Yu-Hsiang, Alexander J. Cooper, Bochuan Teng, Rui B. Chang, Daniel J. Artiga, Heather N. Turner, Eric M. Mulhall, Wenlei Ye, Andrew D. Smith, and Emily R. Liman, "An evolutionarily conserved gene family encodes proton-selective ion channels," *Science* 359, no. 6379 (2018): 1047–1050.

2. Beer is a tricky beverage to brew. It requires at least three steps. The first stage is conversion of the starches in the grains into sugars. One way to do this is through malting, in which the grain is germinated and enzymes from the grain help to convert starches in the grain into simple sugar. Then the barley must be heated and mashed. It is only after these two steps that fermentation can begin. If it begins. In many cases,

fermentation requires an inoculum of some kind, a source of bacteria and yeast akin to a sourdough starter. Compared to beer making, it is much easier to ferment honey into mead. It is easier still to ferment apples into cider or palm fruits or grapes into wine. One scarcely needs to intervene. One only needs to gather the fruit in a receptacle of some kind—the gut of an animal, a hollowed-out gourd, or even a hole in the ground—and wait.

3. We asked Liz if she tasted the fruits. Her response was, "I haven't tasted these—I stopped trying capuchin foods when I ate one that made my mouth go numb. It was perhaps a poor choice to try a berry from a stinging nettle relative."

4. Yeasts are fungi with an unusual lifestyle. Most live on sugar, which they metabolize. Rather than growing through sugar using filaments (the way bread mold colonizes your sandwich), yeasts simply divide. One cell becomes two, which become four and so on. The first problem for yeast in fulfilling this exponential life goal is that sugar is actually very rare in the environment. After yeasts have found, say, a flower, colonized its nectar, and eaten all of the sugar therein, they need to find more sugar. But they don't grow filaments and don't have wings. They can't even get airborne very readily. Yeast cells are fat and heavy. Even if they tried to get airborne they would just flop back down like fat, featherless baby birds. Yeast cells depend on catching a ride with animals to new sources of sugar. They ride bees and wasps from flower to flower and fruit to fruit. Waiting on insects can be a risky endeavor, and so yeasts have evolved the ability to produce aromas that attract insects. When microbiologist Anne Madden was a researcher in Rob's laboratory, she found that most bees and wasps, at any given moment, are unwittingly carrying around yeast cells. Wasps are taxis taking passengers from one sweet bar to the next. The insects aren't entirely duped, though. Just as mammals get a reward of fruit for carrying the seeds of plants, insects that carry yeasts also often get a reward for partnering with yeasts. By producing airborne aromas as they divide and metabolize, yeasts help insects to find sugar sources that they might not otherwise notice. "Sugar over here!" Anne A. Madden, Mary Jane Epps, Tadashi Fukami, Rebecca E. Irwin, John Sheppard, D. Magdalena Sorger, and Robert R. Dunn, "The ecology of insect-yeast relationships and its relevance to human industry," *Proceedings of the Royal Society B: Biological Sciences* 285, no. 1875 (2018): 20172733.

5. Acetaldehyde is then converted to acetaldehyde acetate by acetaldehyde dehydrogenase.

6. As an aside, most of the first fermentations are likely to have been sour due to lactic acid bacteria, but some would have been less sour than others. Fermentations with grapes tend to favor yeast over bacteria. Grapes contain tartaric acid. Most bacteria can't metabolize tartaric acid, but yeasts can. As a result, the spontaneous fermentation of grapes tends to produce a drink that is more alcoholic and less sour than does spontaneous fermentation of malted grains..

7. Archaeologists have long practiced the science of re-creation. It is often hard to know how something might have been done without doing it yourself. In another such re-creation, an archaeologist sought to figure out how Clovis hunters might have used Clovis-tipped spears to kill mastodons or mammoths (or even whether doing so was possible). He took advantage of an elephant cull that was going on at the time in Zimbabwe. Each time an elephant was killed as part of the cull, he would run up to it and fling a spear at its side using an atlatl. It worked. The spear penetrated the elephant's hide with relative ease. Such experiments do not show what happened in the past, but they do certainly help to show what might have been possible. Gary Haynes and Janis Klimowicz, "Recent elephant-carcass utilization as a basis for interpreting mammoth exploitation," *Quaternary International* 359 (2015): 19–37.

8. Fisher's final experiment on this topic was to be a kind of observation. As Fisher reported it to Rob in an email, he "was asked to recover a buried elephant for the Toledo Zoo. The elephant had died 17 years before and had been buried in dense clay and had fermented in the ground over that time, retaining most of its soft tissue as 'pickled' meat, highly acidic and with an even stronger smell than the horse. I didn't try eating the elephant but did do butchery experiments with it over the course of about three days." The result of those experiments was the realization that the elephant's innards had become very, very acidic.

9. If the fermentation of meat is, as Fisher argues, so potentially valuable to predators and omnivores, one might ask why wolves and other carnivores haven't evolved ways to store and ferment meat and, in doing so, favor beneficial microbes and disfavor problem species. They do! Like humans, many wild carnivores often kill more in one hunt than they can eat fresh. Sometimes this is because they kill prey animals far bigger than their stomachs. Sometimes it is because, even after killing as much prey as they can eat, they keep killing more. The red fox (*Vulpes vulpes*) is notorious for killing far more than it seems to need. The fox in the hen house is such a problem because it may kill not just one hen, but dozens. Then what? The truth is, carnivores do store and ferment meat. Or at least some do. After killing more than they need, red foxes store prey items beneath the snow or dirt. They tend to choose sunny spots to ferment their meat. Wolves do the same. Bears bury prey items and then cover them with sphagnum, a plant that disfavors some microbes and favors others. Hyenas put meat in water. In each of these cases, the animals later return to the meat they have stored, after it has fermented, and eat it. They devour their ferments. No one has assessed how these animals choose which meat to eat after it has been stored and fermented. We hypothesize that they preferentially choose acidic and hence sour meat. Unfortunately, we could not find a single study on the abilities of a carnivore (any carnivore) to detect sour tastes and whether those tastes tend to be innately attractive, aversive, or learned. We don't even appear to know the answer very well for domestic dogs. C. C. Smith and O. J. Reichman, "The evolution of food caching

by birds and mammals," *Annual Review of Ecology and Systematics* 15, no. 1 (1984): 329–51; D. F. Sherry, "Food storage by birds and mammals," in *Advances in the Study of Behavior* (Academic Press, 1985), vol. 15, 153–88.

10. We are drawing heavily here on the excellent work of John Speth, another University of Michigan paleoanthropologist. Speth reviews the literature on meat and fish fermentation and collates many of these cultural references. J. D. Speth, "Putrid meat and fish in the Eurasian middle and upper Paleolithic: Are we missing a key part of Neanderthal and modern human diet?" *PaleoAnthropology* (2017): 44–72.

11. It was originally fermented in large wooden barrels, but the advent of canning allowed it to be made (in various containers) at industrial scale and shipped and distributed across Sweden. The tradition in Höga Kusten (The High Coast), where surströmming is considered to have originated, is to make a sandwich of two pieces of buttered thin bread with surströmming and potatoes in between. This sandwich is accompanied by schnapps for good measure. For more of the story of this fascinating food, see Torstein Skåra, Lars Axelsson, Gudmundur Stefánsson, Bo Ekstrand, and Helge Hagen, "Fermented and ripened fish products in the northern European countries," *Journal of Ethnic Foods* 2, no. 1 (2015): 18–24.

12. Fermented Greenlandic shark shares a similar cultural history with these other foods, but is a little different in that the shark cannot be eaten unless it is fermented. The shark meat contains large amounts of urea and trimethyl oxide, which can be poisonous but are made less so by fermentation. Fermentation converts the urea to ammonia, which is not poisonous but, of course, smells like ammonia. Great fondness for particular flavor can be born out of real hardship and yet remain a fondness.

13. The fish sauce was made much in the way that other fermented fish products were made. The Romans impregnated fish with salt (two pints of salt to a peck of fish). The salt was mixed in well and left for a night. Then the salt and fish were put in a ceramic vessel and stored in the sun for months or even as long as a year. The Romans then used the sauce that oozed from the fish as a kind of flavoring called *liquamen* or *garum*. As Tannahill notes, it could be made finer with higher quality fish or shrimp (or a little bit of added wine), but it was just fine made with the ordinary fish, typically sprats, anchovies, horse mackerel, or mackerel. Historians know about the value of garum in the ancient Roman Empire via many documents, but one of our personal favorites is a letter dated to the year 230 CE. It is a letter sent from a Greek man, Arrianus, to his brother Paulus. In it, Arrianus offers some general pleasantries and words of greeting. But it is clear that the actual reason for the letter was something else. Arrianus wanted some fermented fish paste. He wrote, "Send me the fish liver sauce too, whichever you think is good. . . ." The fish liver sauce was garum, the result of a fermentation so good it was worth writing home to ask for.

Garum is no longer eaten as an everyday food (though we were just served fish with garum at a restaurant in Copenhagen, and the new Noma guide to fermentation features a dozen new types of garum, including one made from its grasshoppers). René Redzepi and David Zilber, *The Noma Guide to Fermentation: Including Koji, Kombuchas, Shoyus, Misos, Vinegars, Garums, Lacto-ferments, and Black Fruits and Vegetables* (Artisan Books, 2018).

14. Even as stinkfish and related foods are becoming more appreciated in a small subset of restaurants, they are still often stigmatized in the communities for whom they offer sustenance. Modern Native Americans still struggle with colonial perspectives about the "proper" aroma of food. Sveta Yamin-Pasternak, Andrew Kliskey, Lilian Alessa, Igor Pasternak, Peter Schweitzer, Gary K. Beauchamp, Melissa L. Caldwell, et al., "The rotten renaissance in the Bering Strait: Loving, loathing, and washing the smell of foods with a (re)acquired taste." *Current Anthropology* 55, no. 5 (2014): 619–46.

15. This chapter was improved by Joanna Lambert, Sally Grainger, Li Liu, Michael Kalanty, Paul Breslin, Sveta Yamin-Pasternak, Adam Boethius, Tate Paulette, Jessie Hendy, Daniel Fisher, Torstein Skåra, Emily Liman, Katie Amato, Matthew Booker, Sevgi Sirakova Mutlu, Chad Ludington, John Speth, Amaia Arranz Otaegui, Matthew Carrigan, Daniel Fisher, and Shinya Shoda. Once again, Josh Evans and Kim Wejendorp added a culinary perspective, as did Sandor Katz and, over coffee and very delicious pastries, David Zilber.

Chapter 8. The Art of Cheese

1. Bunyard originally wrote "man" and "manly" rather than "human." However, he did so not to emphasize the gender of the cheese, but instead its humanness. We've altered the text so as to preserve the sentiment, that cheese becomes more like a human through time as it ferments.

2. Jose eventually left Carreña and the cheese caves and moved to the United States, where he studies the *Lactobacillus* bacteria found in cheese and other foods. Manolo stayed with the cheese. Jose, like Rob, works at North Carolina State University.

3. The food eaten by the animals can influence the cheese in many ways. It can influence the energy available for the mother animal and hence the quantities of protein and fat in the milk. It can also influence the flavors of the milk inasmuch as some of the compounds in the plants dairy animals eat can travel into their milk. But the effects can also be more complex. Recently a study in Herbipôle, France, showed that animals that forage more extensively have different microbes on their udders, and on their skin more generally, which leads to different microbes in the milk and, ultimately, different microbes and aromas in the cheeses made from that milk. Marie Frétin, Bruno Martin, Etienne Rifa, Verdier-Metz Isabelle, Dominique Pomiès, Anne

Ferlay, Marie-Christine Montel, and Céline Delbès, "Bacterial community assembly from cow teat skin to ripened cheeses is influenced by grazing systems," *Scientific Reports* 8, no. 1 (2018): 200.

4. As Crystal Louramore-Kirsanova, a medieval Russian historian who also happens to have a fondness for the Benedictine Rules and cheese, pointed out to Rob, these early rules could be very strict. For instance, Crystal notes, Saint Casian's rule book lays out rules for a monk's shoes. "Monks are forbidden to wear shoes, but if 'bodily weakness' compels them, they are to protect their feet with sandals. Monks have to explain their use of the sandals and obtain the Lord's permission. They then have to admit that while they live in this world, 'they cannot be completely set free from care and anxiety about the flesh' and 'always [be] prepared for preaching the peace of the gospel' as the Lord permitted them to make use of the sandals in the first place."

5. Or with some kind of ladle or other tool, depending on the place, time, and cheese.

6. Gouda, for instance, has a unique flavor some have described as a mix of chocolate, banana, and sweat, due to the presence of methylproponal (the chocolate and banana) and butyric acid (the sweat).

7. Cheesemakers in central Asia actually make their cheese using an approach very similar to that employed in curing meat; they dry their cheese in the sun and add salt as it dries.

8. Both these cheeses and washed-rind cheeses were probably made by some farmers before the monasteries came into existence, perhaps even in Roman times. They would have been made in small batches, though, and so are little documented. The monks both helped to preserve these local cheeses and contributed new varieties. Because the monks ultimately get credit for everything associated with the monastery (whether they made the food or were just given the food as tithe), it is often difficult to tell the difference between these scenarios.

9. In these and other aged cheeses, a kind of succession occurs through time. Lactic acid bacteria are the first colonizers. The *Penicillium* fungi then metabolize the lactic acid produced by these bacteria. Once they have, other bacteria that are not tolerant of lactic acid and are more characteristic of human skin move in. This succession is predictable, or at least it is predictable if everything is done right.

10. A secondary benefit was that the constant brining helped to keep the cheese flies at bay.

11. Indeed, these cheeses played much of the role of meat for the monks, with up to 30 percent protein and 30 percent fat.

12. *Munster* means monastery in the dialect of German spoken in the Alsace region of France in which the cheese is made.

13. Stephane and Jeni are married; he is French and fond of Époisses; she is American and, well, not. Nor are Stephane and Jeni alone in their difference of

opinion. More than one cheese lover, in reading this paragraph, went on to describe the special measures they take to prevent their spouse from smelling their favorite stinky cheese (double layers of Tupperware, separate refrigerators, and cellars, for example). Art can be in the nose of the beholder.

14. Ben Wolfe, Jose Bruno-Bárcena, Matthew Booker, Sevgi Sirakova Mutlu, Chad Ludington, Benoit Guénard, Jessica Hendy, Michael Dunn, Aminah Al-Attas Bradford, Shinya Shoda, Tate Paulette, Matthäus Rest, and Heather Paxson all read and improved this chapter. Josh Evans, David Asher, Sandor Katz, and Kim Wejendorp added a culinary perspective.

Chapter 9. Dinner Makes Us Human

1. Much of the English-language world of food was defined by French writers and the French language. We have a gourmet banquet at which we eat hors d'oeuvres, sip a consommé, followed by an entrée in which vegetables have been sautéed and accompanied by a pâté. We then drink from a carafe of wine. So in talking about a supper that is grand, we have no word to use to describe it but to call it a banquet or fête. It was a fête.

2. It seemed like we'd had the sort of night one has just once in a lifetime, and then, four nights later, in another French town, this time Limeuil, we had a very similar evening.

3. Roman was somewhere between a roasting pig, a keg of beer, and the host's twin daughters, who appeared to be, for reasons we can't recall, auctioning items from their house that they hoped no one would miss. The dinner was a going-away party for friends at a large house at the edge of a park. That Roman happened to work on chimpanzees and be at the same party we were attending is a testament to the relatively small size of Leipzig, the relatively large number of chimpanzee researchers that live there, and the extent to which social groups in Leipzig, like most cities, tend to be non-random. The guests were mostly parents of children at Leipzig International School.

4. They sometimes even go out of their way to do so. For example, Wittig observed a case in Uganda involving an alpha male, Nick, from the Sonso community. As Wittig described via email, Nick was "under pressure from the second in rank, Bwoba, and needed an ally. One day Nick hunted and caught a colobus monkey, then carried the dead monkey without feeding on it for over one kilometer, pant-hooting and searching for Zefa, a powerful male he wanted to befriend who had not taken part in the hunt. Only after fifteen minutes, when Nick had found Zefa, did he rip the carcass in two pieces. Nick then gave Zefa the larger part, and they ate together (you take the head, I'll take the hand, you take the brain, I'll take the leg, and let's be

friends). On a more recent trip, Liran Samuni saw something similar when watching chimpanzees gather jackfruits.

5. Samuni also found that male chimpanzees, while hunting together, also receive a bump in oxytocin. Eating together and trying to get food together have similarly pleasing effects. Liran Samuni, Anna Preis, Tobias Deschner, Catherine Crockford, and Roman M. Wittig, "Reward of labor coordination and hunting success in wild chimpanzees," *Communications Biology* 1, no. 1 (2018): 1–9.

6. For example, we didn't tell the story, told to us by a Danish ornithologist, about the night that the Danish opera was rehearsing a Wagner performance in the Frederiksberg Park near the zoo. The performance proved too much for the okapi, who died in panic. Not a group to allow an animal to go to waste, the scientists skinned the okapi and preserved its scientifically important parts for future study. They then cooked and devoured the rest, which is said to have been delicious. Or rather, we almost didn't tell that story. Nor did we tell the story about the . . . well, anyway, you get the idea.

7. Matthew Booker, Ammie Kalan, Chad Ludington, Maureen McCarthy, Roman Wittig, Liran Samuni, Athena Aktipis, James Rives, August Sanchez Dunn, and Olivia Sanchez Dunn all read and commented upon this chapter. Once again, Josh Evans and Kim Wejendorp added a culinary perspective. Lisa Raschke and Lynne Trautwein provided insights here and throughout the book.

REFERENCES

[1] Hsiang Ju Lin and Tsuifeng Lin, *The Art of Chinese Cuisine* (Tuttle Publishing, 1996).

[2] Jean Anthelme Brillat-Savarin, *La physiologie du goût* [1825], ed. Jean-François Revel (Paris: Flammarion, 1982), 19.

[3] Richard Stevenson, *The Psychology of Flavour* (Oxford University Press, 2010).

[4] Gordon M. Shepherd, *Neurogastronomy: How the Brain Creates Flavor and Why It Matters* (Columbia University Press, 2011).

[5] Charles Spence, *Gastrophysics: The New Science of Eating* (Penguin UK, 2017); Ole Mouritsen and Klavs Styrbæk, *Mouthfeel: How Texture Makes Taste*, translated by Mariela Johansen (Columbia University Press, 2017).

[6] Paul A. S. Breslin, "An evolutionary perspective on food and human taste," *Current Biology* 23, no. 9 (2013): R409–18.

[7] Jonathan Silvertown, *Dinner with Darwin: Food, Drink, and Evolution* (University of Chicago Press, 2017).

[8] Ken'ichi Ikeda, "On a new seasoning," *Journal of the Tokyo Chemical Society* 30 (1909): 820–36. The paper appears to have been first referenced in an English-language paper in 1966.

[9] Jonathan P. Benstead, James M. Hood, Nathan V. Whelan, Michael R. Kendrick, Daniel Nelson, Amanda F. Hanninen, and Lee M. Demi, "Coupling of dietary phosphorus and growth across diverse fish taxa: A meta-analysis of experimental aquaculture studies," *Ecology* 95, no. 10 (2014): 2768–77.

[10] Stuart A. McCaughey, Barbara K. Giza, and Michael G. Tordoff, "Taste and acceptance of pyrophosphates by rats and mice," *American Journal of Physiology Regulatory Integrative and Comparative Physiology* 292 (2007): R2159–67.

[11] D. J. Holcombe, David A. Roland, and Robert H. Harms, "The ability of hens to regulate phosphorus intake when offered diets containing different levels of phosphorus," *Poultry Science* 55 (1976): 308–17; G. M. Siu, Mary Hadley, and Harold H. Draper, "Self-regulation of phosphate intake by growing rats," *Journal of Nutrition* 111, no. 9 (1981): 1681–85; Juan J. Villalba, Frederick D. Provenza,

Jeffery O. Hall, and C. Peterson, "Phosphorus appetite in sheep: Dissociating taste from postingestive effects," *Journal of Animal Science* 84, no. 8 (2006): 2213–23.

[12] Michael G. Tordoff, "Phosphorus taste involves T1R2 and T1R3," *Chemical Senses* 42, no. 5 (2017): 425–33; Michael G. Tordoff, Laura K. Alarcón, Sitaram Valmeki, and Peihua Jiang, "T1R3: A human calcium taste receptor," *Scientific Reports* 2 (2012): 496.

[13] Diane W. Davidson, Steven C. Cook, Roy R. Snelling, and Tock H. Chua, "Explaining the abundance of ants in lowland tropical rainforest canopies," *Science* 300, no. 5621 (2003): 969–72.

[14] Anne Fischer, Yoav Gilad, Orna Man, and Svante Pääbo, "Evolution of bitter taste receptors in humans and apes," *Molecular Biology and Evolution* 22, no. 3 (2004): 432–36.

[15] Xia Li, Weihua Li, Hong Wang, Douglas L. Bayley, Jie Cao, Danielle R. Reed, Alexander A. Bachmanov, Liquan Huang, Véronique Legrand-Defretin, Gary K. Beauchamp, and Joseph G. Brand, "Cats lack a sweet taste receptor," *Journal of Nutrition* 136, no. 7 (2006): 1932S–1934S; Peihua Jiang, Jesusa Josue, Xia Li, Dieter Glaser, Weihua Li, Joseph G. Brand, Robert F. Margolskee, Danielle R. Reed, and Gary K. Beauchamp, "Major taste loss in carnivorous mammals," *Proceedings of the National Academy of Sciences* 109, no. 13 (2012): 4956–61.

[16] Peihua Jiang, Jesusa Josue, Xia Li, Dieter Glaser, Weihua Li, Joseph G. Brand, Robert F. Margolskee, Danielle R. Reed, and Gary K. Beauchamp, "Major taste loss in carnivorous mammals," *Proceedings of the National Academy of Sciences* 109, no. 13 (2012): 4956–61.

[17] Zhao Huabin, Jian-Rong Yang, Huailiang Xu, and Jianzhi Zhang, "Pseudogenization of the umami taste receptor gene Tas1r1 in the giant panda coincided with its dietary switch to bamboo," *Molecular Biology and Evolution* 27, no. 12 (2010): 2669–73.

[18] Peihua Jiang, Jesusa Josue-Almqvist, Xuelin Jin, Xia Li, Joseph G. Brand, Robert F. Margolskee, Danielle R. Reed, and Gary K. Beauchamp, "The bamboo-eating giant panda (*Ailuropoda melanoleuca*) has a sweet tooth: Behavioral and molecular responses to compounds that taste sweet to humans," *PloS One* 9, no. 3 (2014).

[19] Shancen Zhao, Pingping Zheng, Shanshan Dong, Xiangjiang Zhan, Qi Wu, Xiaosen Guo, Yibo Hu et al., "Whole-genome sequencing of giant pandas provides insights into demographic history and local adaptation," *Nature Genetics* 45, no. 1 (2013): 67.

[20] Maude W. Baldwin, Yasuka Toda, Tomoya Nakagita, Mary J. O'Connell, Kirk C. Klasing, Takumi Misaka, Scott V. Edwards, and Stephen D. Liberles, "Evolution of sweet taste perception in hummingbirds by transformation of the ancestral umami receptor," *Science* 345, no. 6199 (2014): 929–33.

[21] Ricardo A. Ojeda, Carlos E. Borghi, Gabriela B. Diaz, Stella M. Giannoni, Michael A. Mares, and Janet K. Braun, "Evolutionary convergence of the highly adapted desert rodent *Tympanoctomys barrerae* (Octodontidae)," *Journal of Arid Environments* 41, no. 4 (1999): 443–52.

[22] David R. Pilbeam and Daniel E. Lieberman, "Reconstructing the last common ancestor of chimpanzees and humans," In *Chimpanzees and Human Evolution*, ed. M. N. Muller (Harvard University Press, 2017), 22–141.

[23] Charles Darwin, *The Descent of Man, and Selection in Relation to Sex* (John Murray, 1888).

[24] Jane Goodall, "Tool-using and aimed throwing in a community of free-living chimpanzees," *Nature* 201, no. 4926 (1964): 1264.

[25] Christophe Boesch, Ammie K. Kalan, Anthony Agbor, Mimi Arandjelovic, Paula Dieguez, Vincent Lapeyre, and Hjalmar S. Kühl, "Chimpanzees routinely fish for algae with tools during the dry season in Bakoun, Guinea," *American Journal of Primatology* 79, no. 3 (2017): e22613.

[26] Hitonaru Nishie, "Natural history of *Camponotus* ant-fishing by the M group chimpanzees at the Mahale Mountains National Park, Tanzania," *Primates* 52, no. 4 (2011): 329.

[27] Christophe Boesch, *Wild Cultures: A Comparison between Chimpanzee and Human Cultures* (Cambridge University Press, 2012).

[28] Solomon H. Katz, "An evolutionary theory of cuisine," *Human Nature* 1, no. 3 (1990): 233–59.

[29] David R. Pilbeam and Daniel E. Lieberman, "Reconstructing the last common ancestor of chimpanzees and humans," in *Chimpanzees and Human Evolution*, ed. M. N. Muller (Harvard University Press, 2017), 22–141.

[30] T. Jonathan Davies, Barnabas H. Daru, Bezeng S. Bezeng, Tristan Charles-Dominique, Gareth P. Hempson, Ronny M. Kabongo, Olivier Maurin, A. Muthama Muasya, Michelle van der Bank, and William J. Bond, "Savanna tree evolutionary ages inform the reconstruction of the paleoenvironment of our hominin ancestors," *Scientific Reports* 10, no. 1 (2020): 1–8.

[31] Jill D. Pruetz and Nicole M. Herzog, "Savanna chimpanzees at Fongoli, Senegal, navigate a fire landscape," *Current Anthropology* 58, no. S16 (2017): S337–50.

[32] Thomas S. Kraft and Vivek V. Venkataraman, "Could plant extracts have enabled hominins to acquire honey before the control of fire?" *Journal of Human Evolution* 85 (2015): 65–74; Lidio Cipriani, ed., *The Andaman Islanders* (Weidenfeld and Nicolson, 1966).

[33] Christophe Boesch, Ammie K. Kalan, Anthony Agbor, Mimi Arandjelovic, Paula Dieguez, Vincent Lapeyre, and Hjalmar S. Kühl, "Chimpanzees routinely fish for algae with tools during the dry season in Bakoun, Guinea," *American Journal of Primatology* 79, no. 3 (2017): e22613.

[34] Kathelijne Koops, Richard W. Wrangham, Neil Cumberlidge, Maegan A. Fitzgerald, Kelly L. van Leeuwen, Jessica M. Rothman, and Tetsuro Matsuzawa, "Crab-fishing by chimpanzees in the Nimba Mountains, Guinea," *Journal of Human Evolution* 133 (2019): 230–41.

[35] William H. Kimbel, Robert C. Walter, Donald C. Johanson, Kaye E. Reed, James L. Aronson, Zelalem Assefa, Curtis W. Marean, Gerald G. Eck, René Bobe, Erella Hovers, Yoel Zvi Rak, Carl Vondra, Tesfaye Yemane, D. York, Yanchao Chen, Norman M. Evensen, and Patrick E. Smith, "Late Pliocene *Homo* and Oldowan tools from the Hadar formation (Kada Hadar member), Ethiopia," in R. L. Chiochon and J. G. Fleagle, eds., *The Human Evolution Source Book* (Routledge, 2016).

[36] Melissa J. Remis, "Food preferences among captive western gorillas (*Gorilla gorilla gorilla*) and chimpanzees (*Pan troglodytes*)," *International Journal of Primatology* 23, no. 2 (2002): 231–49.

[37] Victoria Wobber, Brian Hare, and Richard Wrangham. "Great apes prefer cooked food," *Journal of Human Evolution* 55, no. 2 (2008): 340–48.

[38] Daniel Lieberman, *The Story of the Human Body: Evolution, Health, and Disease* (Vintage, 2014).

[39] Toshisada Nishida and Mariko Hiraiwa, "Natural history of a tool-using behavior by wild chimpanzees in feeding upon wood-boring ants," *Journal of Human Evolution* 11, no. 1 (1982): 73–99.

[40] Matthew R. McLennan, "Diet and feeding ecology of chimpanzees (*Pan troglodytes*) in Bulindi, Uganda: Foraging strategies at the forest–farm interface," *International Journal of Primatology* 34, no. 3 (2013): 585–614.

[41] Matthew R. McLennan, Georgia A. Lorenti, Tom Sabiiti, and Massimo Bardi, "Forest fragments become farmland: Dietary response of wild chimpanzees (*Pan troglodytes*) to fast-changing anthropogenic landscapes," *American Journal of Primatology* 82, no. 4 (2020): e23090.

[42] Julia Colette Berbesque and Frank W. Marlowe, "Sex differences in food preferences of Hadza hunter-gatherers," *Evolutionary Psychology* 7, no. 4 (2009): 147470490900700409.

[43] Hsiang Ju Lin and Tsuifeng Lin, *The Art of Chinese Cuisine* (Tuttle, 1996).

[44] Chris Organ, Charles L. Nunn, Zarin Machanda and Richard W. Wrangham, "Phylogenetic rate shifts in feeding time during the evolution of *Homo*," *Proceedings of the National Academy of Sciences* 108, no. 35 (2011): 14555–59.

[45] Victoria Wobber, Brian Hare, and Richard Wrangham, "Great apes prefer cooked food," *Journal of Human Evolution* 55, no. 2 (2008): 340–48; Felix Warneken and Alexandra G. Rosati, "Cognitive capacities for cooking in chimpanzees," *Proceedings of the Royal Society B: Biological Sciences* 282, no. 1809 (2015): 20150229.

[46] Peter S. Ungar, Frederick E. Grine, and Mark F. Teaford, "Diet in early *Homo*: A review of the evidence and a new model of adaptive versatility," *Annual Review of Anthropology* 35 (2006): 209–28.

[47] Ruth Blasco, Jordi Rosell, M. Arilla, Antoni Margalida, D. Villalba, Avi Gopher, and Ran Barkai, "Bone marrow storage and delayed consumption at Middle Pleistocene Qesem Cave, Israel (420 to 200 ka)," *Science Advances* 5, no. 10 (2019): eaav9822.

[48] Kohei Fujikura, "Multiple loss-of-function variants of taste receptors in modern humans," *Scientific Reports* 5 (2015): 12349.

[49] Thomas D. Bruns, Robert Fogel, Thomas J. White, and Jeffrey D. Palmer, "Accelerated evolution of a false-truffle from a mushroom ancestor," *Nature* 339, no. 6220 (1989): 140–42.

[50] Daniel S. Heckman, David M. Geiser, Brooke R. Eidell, Rebecca L. Stauffer, Natalie L. Kardos, and S. Blair Hedges, "Molecular evidence for the early colonization of land by fungi and plants," *Science* 293, no. 5532 (2001): 1129–33.

[51] Eva Streiblová, Hana Gryndlerova, and Milan Gryndler, "Truffle brûlé: An efficient fungal life strategy," *FEMS Microbiology Ecology* 80, no. 1 (2012): 1–8.

[52] Jeffrey B. Rosen, Arun Asok, and Trisha Chakraborty, "The smell of fear: Innate threat of 2, 5-dihydro-2, 4, 5-trimethylthiazoline, a single molecule component of a predator odor," *Frontiers in Neuroscience* 9 (2015): 292.

[53] Ken Murata, Shigeyuki Tamogami, Masamichi Itou, Yasutaka Ohkubo, Yoshihiro Wakabayashi, Hidenori Watanabe, Hiroaki Okamura, Yukari Takeuchi, and Yuji Mori, "Identification of an olfactory signal molecule that activates the central regulator of reproduction in goats," *Current Biology* 24, no. 6 (2014): 681–86.

[54] David R. Kelly, "When is a butterfly like an elephant?" *Chemistry and Biology* 3, no. 8 (1996): 595–602.

[55] Thierry Talou, Antoine Gaset, Michel Delmas, Michel Kulifaj, and Charles Montant, "Dimethyl sulphide: The secret for black truffle hunting by animals?" *Mycological Research* 94, no. 2 (1990): 277–78.

[56] Frido Welker, Jazmín Ramos-Madrigal, Petra Gutenbrunner, Meaghan Mackie, Shivani Tiwary, Rosa Rakownikow Jersie-Christensen, Cristina Chiva, Marc R. Dickinson, Martin Kuhlwilm, Marc de Manuel, Pere Gelabert, María Martinón-Torres, Ann Margvelashvili, Juan Luis Arsuaga, Eudald Carbonell, Tomas Marques-Bonet, Kirsty Penkman, Eduard Sabidó, Jürgen Cox, Jesper V. Olsen, David Lordkipanidze, Fernando Racimo, Carles Lalueza-Fox, José María Bermúdez de Castro, Eske Willerslev, and Enrico Cappellini, "The dental proteome of *Homo antecessor*," *Nature* 580 (2020): 1–4.

[57]. Paul Mellars and Jennifer C. French, "Tenfold population increase in Western Europe at the neandertal-to-modern human transition," *Science* 333, no. 6042 (2011): 623–27.

[58] Neil Shubin, *Your Inner Fish: A Journey into the 3.5-Billion-Year History of the Human Body* (Vintage, 2008).

[59] Yoshihito Niimura, "Olfactory receptor multigene family in vertebrates: From the viewpoint of evolutionary genomics," *Current Genomics* 13, no. 2 (2012): 103–14.

[60] Gordon M. Shepherd, *Neurogastronomy: How the Brain Creates Flavor and Why It Matters* (Columbia University Press, 2011).

[61] Katherine A. Houpt and Sharon L. Smith, "Taste preferences and their relation to obesity in dogs and cats," *Canadian Veterinary Journal* 22, no. 4 (1981): 77.

[62] Yoav Gilad, Victor Wiebe, Molly Przeworski, Doron Lancet, and Svante Pääbo, "Loss of olfactory receptor genes coincides with the acquisition of full trichromatic vision in primates," *PLoS Biology* 2, no. 1 (2004): e5; Yoshihito Niimura, Atsushi Matsui and Kazushige Touhara, "Acceleration of olfactory receptor gene loss in primate evolution: Possible link to anatomical change in sensory systems and dietary transition," *Molecular Biology and Evolution* 35, no. 6 (2018): 1437–50.

[63] David Zwicker, Rodolfo Ostilla-Mónico, Daniel E. Lieberman, and Michael P. Brenner, "Physical and geometric constraints shape the labyrinth-like nasal cavity," *Proceedings of the National Academy of Sciences* 115, no. 12 (2018): 2936–41.

[64] Luca Pozzi, Jason A. Hodgson, Andrew S. Burrell, Kirstin N. Sterner, Ryan L. Raaum, and Todd R. Disotell, "Primate phylogenetic relationships and divergence," *Molecular Phylogenetics and Evolution* 75 (2014): 165–83.

[65] Daniel E. Lieberman, "How the unique configuration of the human head may enhance flavor perception capabilities: An evolutionary perspective," *Frontiers in Integrative Neuroscience Conference Abstract: Science of Human Flavor Perception* (2015): doi: 10.3389/conf.fnint.2015.03.00003.

[66] Robert D. Martin, *Primate Origins and Evolution* (Chapman and Hall, 1990).

[67] Daniel E. Lieberman, "How the unique configuration of the human head may enhance flavor perception capabilities: An evolutionary perspective," *Frontiers in Integrative Neuroscience Conference Abstract: Science of Human Flavor Perception* (2015): doi: 10.3389/conf.fnint.2015.03.00003.

[68] Susann Jänig, Brigitte M. Weiß, and Anja Widdig, "Comparing the sniffing behavior of great apes," *American Journal of Primatology* 80, no. 6 (2018): e22872.

[69] Arthur W. Proetz, "The Semon Lecture: Respiratory air currents and their clinical aspects," *Journal of Laryngology and Otology* 67, no. 1 (1953): 1–27.

[70] Timothy B. Rowe and Gordon M. Shepherd, "Role of ortho-retronasal olfaction in mammalian cortical evolution," *Journal of Comparative Neurology* 524, no. 3 (2016): 471–95.

[71] Harold McGee, *Curious Cook: More Kitchen Science and Lore* (North Point, 1990).

[72] Andreas Keller and Leslie B. Vosshall, "Olfactory perception of chemically diverse molecules," *BMC Neuroscience* 17, no. 1 (2016): 55.

[73] Harold McGee, *The Curious Cook: More Kitchen Science and Lore* (Wiley, 1992).

[74] Brian Farneti, Iuliia Khomenko, Marcella Grisenti, Matteo Ajelli, Emanuela Betta, Alberto Alarcon Algarra, Luca Cappellin, Eugenio Aprea, Flavia Gasperi, Franco Biasioli, and Lara Giongo, "Exploring blueberry aroma complexity by chromatographic and direct-injection spectrometric techniques," *Frontiers in Plant Science* 8 (2017): 617.

[75] Gordon M. Shepherd, *Neuroenology: How the Brain Creates the Taste of Wine* (Columbia University Press, 2016).

[76] Yukio Takahata, Mariko Hiraiwa-Hasegawa, Hiroyuki Takasaki, and Ramadhani Nyundo, "Newly acquired feeding habits among the chimpanzees of the Mahale Mountains National Park, Tanzania," *Human Evolution* 1, no. 3 (1986): 277–84.

[77] Ibid.

[78] Ciprian F. Ardelean, Lorena Becerra-Valdivia, Mikkel Winther Pedersen, Jean-Luc Schwenninger, Charles G. Oviatt, Juan I. Macías-Quintero, Joaquin Arroyo-Cabrales, Martin Sikora, et al., "Evidence of human occupation in Mexico around the Last Glacial Maximum," *Nature* 584, no. 7819 (2020): 87–92.

[79] M. Thomas P. Gilbert, Dennis L. Jenkins, Anders Götherstrom, Nuria Naveran, Juan J. Sanchez, Michael Hofreiter, Philip Francis Thomsen, et al., "DNA from pre-Clovis human coprolites in Oregon, North America," *Science* 320, no. 5877 (2008): 786–89; Lorena Becerra-Valdivia and Thomas Higham, "The timing and effect of the earliest human arrivals in North America," *Nature* 584 (2020): 1–5.

[80] Michael R. Waters, "Late Pleistocene exploration and settlement of the Americas by modern humans," *Science* 365, no. 6449 (2019): eaat5447.

[81] Michael R. Waters, Thomas W. Stafford, H. Gregory McDonald, Carl Gustafson, Morten Rasmussen, Enrico Cappellini, Jesper V. Olsen, et al., "Pre-Clovis mastodon hunting 13,800 years ago at the Manis site, Washington," *Science* 334, no. 6054 (2011): 351–53.

[82] Michael R. Waters, "Late Pleistocene exploration and settlement of the Americas by modern humans," *Science* 365, no. 6449 (2019): eaat5447.

[83] Gary Haynes and Jarod M. Hutson, "Clovis-era subsistence: Regional variability, continental patterning," in *Paleoamerican Odyssey*, ed. K. E. Graf, C. V. Ketron, and M. R. Waters (Texas A&M University Press, 2014), 293–309.

[84] Joseph A. M. Gingerich, "Down to seeds and stones: A new look at the subsistence remains from Shawnee-Minisink," *American Antiquity* 76, no. 1 (2011): 127–44.

[85] Klervia Jaouen, Michael P. Richards, Adeline Le Cabec, Frido Welker, William Rendu, Jean-Jacques Hublin, Marie Soressi, and Sahra Talamo, "Exceptionally

high δ15N values in collagen single amino acids confirm Neandertals as high-trophic level carnivores," *Proceedings of the National Academy of Sciences* 116, no. 11 (2019): 4928–33.

[86] Michael Chazan, "Toward a long prehistory of fire," *Current Anthropology* 58, no. S16 (2017): S351–59; Alianda M. Cornélio, Ruben E. de Bittencourt-Navarrete, Ricardo de Bittencourt Brum, Claudio M. Queiroz, and Marcos R. Costa, "Human brain expansion during evolution is independent of fire control and cooking," *Frontiers in Neuroscience* 10 (2016): 167.

[87] Alston V. Thoms, "Rocks of ages: Propagation of hot-rock cookery in western North America," *Journal of Archaeological Science* 36, no. 3 (2009): 573–91.

[88] Paul S. Martin, "The Discovery of America: The first Americans may have swept the Western Hemisphere and decimated its fauna within 1000 years," *Science* 179, no. 4077 (1973): 969–74.

[89] Lenore Newman, *Lost Feast: Culinary Extinction and the Future of Food* (ECW Press, 2019).

[90] Henry Nicholls, "Digging for dodo," *Nature* 443 (2006): 138.

[91] Julian P. Hume and Michael Walters, *Extinct Birds* (A & C Black Poyser Imprint, 2012).

[92] Agnes Gault, Yves Meinard, and Franck Courchamp, "Consumers' taste for rarity drives sturgeons to extinction," *Conservation Letters* 1, no. 5 (2008): 199–207.

[93] David P. Watts and Sylvia J. Amsler, "Chimpanzee-red colobus encounter rates show a red colobus population decline associated with predation by chimpanzees at Ngogo," *American Journal of Primatology* 75, no. 9 (2013): 927–37.

[94] Jacquelyn L. Gill, John W. Williams, Stephen T. Jackson, Katherine B. Lininger, and Guy S. Robinson, "Pleistocene megafaunal collapse, novel plant communities, and enhanced fire regimes in North America," *Science* 326, no. 5956 (2009): 1100–1103; Jacquelyn L. Gill, "Ecological impacts of the late Quaternary mega-herbivore extinctions," *New Phytologist* 201, no. 4 (2014): 1163–69.

[95] John D. Speth, *Paleoanthropology and Archaeology of Big-Game Hunting* (Springer, 2012).

[96] Baron Pineda, "Miskito and Misumalpan languages," in *Encyclopedia of Linguistics*, ed. Philipp Strazny (Francis & Taylor Books, 2005).

[97] Jeremy M. Koster, Jennie J. Hodgen, Maria D. Venegas, and Toni J. Copeland, "Is meat flavor a factor in hunters' prey choice decisions?" *Human Nature* 21, no. 3 (2010): 219–42.

[98] Michael D. Cannon and David J. Meltzer, "Explaining variability in Early Paleo-indian foraging," *Quaternary International* 191, no. 1 (2008): 5–17.

[99] Mark Borchert, Frank W. Davis, and Jason Kreitler, "Carnivore use of an avocado orchard in southern California," *California Fish and Game* 94, no. 2 (2008): 61–74.

[100] Tim M. Blackburn and Bradford A. Hawkins, "Bergmann's rule and the mammal fauna of northern North America," *Ecography* 27, no. 6 (2004): 715–24.

[101] Katherine A. Houpt and Sharon L. Smith, "Taste preferences and their relation to obesity in dogs and cats," *Canadian Veterinary Journal* 22, no. 4 (1981): 77.

[102] S. D. Shackelford, J. O. Reagan, Keith D. Haydon, and Markus F. Miller, "Effects of feeding elevated levels of monounsaturated fats to growing-finishing swine on acceptability of boneless hams," *Journal of Food Science* 55, no. 6 (1990): 1485–87.

[103] As translated in *The Food Lover's Anthology* (The Bodleian Anthology: A Literary Compendium, compiled by Peter Hunt, Bodleian Library Publishing, 2014).

[104] Diana Noyce, "Charles Darwin, the Gourmet Traveler," *Gastronomica: The Journal of Food and Culture* 12, no. 2 (2012): 45–52.

[105] Belarmino C. da Silva Neto, André Luiz Borba do Nascimento, Nicola Schiel, Rômulo Romeu Nóbrega Alves, Antonio Souto, and Ulysses Paulino Albuquerque, "Assessment of the hunting of mammals using local ecological knowledge: An example from the Brazilian semiarid region," *Environment, Development and Sustainability* 19, no. 5 (2017): 1795–1813.

[106] Sophie D. Coe, *America's First Cuisines* (University of Texas Press, 2015).

[107] Gary Haynes and Jarod M. Hutson, "Clovis-era subsistence: Regional variability, continental patterning," *Paleoamerican Odyssey* (2013): 293–309.

[108] Laura T. Buck, J. Colette Berbesque, Brian M. Wood, and Chris B. Stringer, "Tropical forager gastrophagy and its implications for extinct hominin diets," *Journal of Archaeological Science: Reports* 5 (2016): 672–79.

[109] Hagar Reshef and Ran Barkai, "A taste of an elephant: The probable role of elephant meat in Paleolithic diet preferences," *Quaternary International* 379 (2015): 28–34.

[110] George E. Konidaris, Athanassios Athanassiou, Vangelis Tourloukis, Nicholas Thompson, Domenico Giusti, Eleni Panagopoulou, and Katerina Harvati, "The skeleton of a straight-tusked elephant (*Palaeoloxodon antiquus*) and other large mammals from the Middle Pleistocene butchering locality Marathousa 1 (Megalopolis Basin, Greece): Preliminary results," *Quaternary International* 497 (2018): 65–84.

[111] Biancamaria Aranguren, Stefano Grimaldi, Marco Benvenuti, Chiara Capalbo, Floriano Cavanna, Fabio Cavulli, Francesco Ciani, et al., "Poggetti Vecchi (Tuscany, Italy): A late Middle Pleistocene case of human–elephant interaction," *Journal of Human Evolution* 133 (2019): 32–60.

[112] Jeffrey J. Saunders and Edward B. Daeschler, "Descriptive analyses and taphonomical observations of culturally-modified mammoths excavated at 'The Gravel Pit,' near Clovis, New Mexico in 1936," *Proceedings of the Academy of Natural Sciences of Philadelphia* (1994): 1–28.

[113] Omer Nevo and Eckhard W. Heymann, "Led by the nose: Olfaction in primate feeding ecology," *Evolutionary Anthropology: Issues, News, and Reviews* 24, no. 4 (2015): 137–48.

[114] H. Martin Schaefer, Alfredo Valido, and Pedro Jordano, "Birds see the true colours of fruits to live off the fat of the land," *Proceedings of the Royal Society B: Biological Sciences* 281, no. 1777 (2014): 20132516.

[115] Kim Valenta and Omer Nevo, "The dispersal syndrome hypothesis: How animals shaped fruit traits, and how they did not," *Functional Ecology* 34, no. 6 (2020): 1158–69.

[116] Daniel H. Janzen, "Why fruits rot, seeds mold, and meat spoils," *American Naturalist* 111, no. 980 (1977): 691–713.

[117] Daniel H. Janzen, "Why tropical trees have rotten cores," *Biotropica* 8 (1976): 110–12.

[118] Daniel H. Janzen, "Herbivores and the number of tree species in tropical forests," *American Naturalist* 104, no. 940 (1970): 501–28.

[119] Daniel H. Janzen and Paul S. Martin, "Neotropical anachronisms: The fruits the gomphotheres ate," *Science* 215, no. 4528 (1982): 19–27.

[120] Guadalupe Sanchez, Vance T. Holliday, Edmund P. Gaines, Joaquín Arroyo-Cabrales, Natalia Martínez-Tagüeña, Andrew Kowler, Todd Lange, Gregory W. L. Hodgins, Susan M. Mentzer, and Ismael Sanchez-Morales, "Human (Clovis)–gomphothere (*Cuvieronius* sp.) association~ 13,390 calibrated yBP in Sonora, Mexico," *Proceedings of the National Academy of Sciences* 111, no. 30 (2014): 10972–77.

[121] Connie Barlow, *The Ghosts of Evolution: Nonsensical Fruit, Missing Partners, and Other Ecological Anachronisms* (Basic Books, 2008).

[122] Daniel H. Janzen, "How and why horses open *Crescentia alata* fruits," *Biotropica* (1982): 149–52.

[123] Guillermo Blanco, Jose Luis Tella, Fernando Hiraldo, and José Antonio Díaz-Luque, "Multiple external seed dispersers challenge the megafaunal syndrome anachronism and the surrogate ecological function of livestock," *Frontiers in Ecology and Evolution* 7 (2019): 328.

[124] Mauro Galetti, Roger Guevara, Marina C. Côrtes, Rodrigo Fadini, Sandro Von Matter, Abraão B. Leite, Fábio Labecca, T. Ribeiro, C. S. Carvalho, R. G. Collevatti, and M. M. Pires, "Functional extinction of birds drives rapid evolutionary changes in seed size," *Science* 340, no. 6136 (2013): 1086–90.

[125] Renske E. Onstein, William J. Baker, Thomas L. P. Couvreur, Søren Faurby, Leonel Herrera-Alsina, Jens-Christian Svenning, and W. Daniel Kissling, "To adapt or go extinct? The fate of megafaunal palm fruits under past global change," *Proceedings of the Royal Society B: Biological Sciences* 285, no. 1880 (2018): 20180882.

[126] David N. Zaya and Henry F. Howe, "The anomalous Kentucky coffeetree: Megafaunal fruit sinking to extinction?" *Oecologia* 161, no. 2 (2009): 221–26.

[127] Robert J. Warren, "Ghosts of cultivation past-Native American dispersal legacy persists in tree distribution," *PloS One* 11, no. 3 (2016).

[128] Maarten Van Zonneveld, Nerea Larranaga, Benjamin Blonder, Lidio Coradin, José I. Hormaza, and Danny Hunter, "Human diets drive range expansion of megafauna-dispersed fruit species," *Proceedings of the National Academy of Sciences* 115, no. 13 (2018): 3326–31.

[129] Allen Holmberg, "Cooking and eating among the Siriono of Bolivia," in Jessica Kuper, ed., *The Anthropologists' Cookbook* (Routledge, 1997).

[130] Napoleon A. Chagnon, *The Yanomamo* (Nelson Education, 2012).

[131] S. J. McNaughton and J. L. Tarrants, "Grass leaf silicification: Natural selection for an inducible defense against herbivores," *Proceedings of the National Academy of Sciences* 80, no. 3 (1983): 790–91.

[132] Brian D. Farrell, David E. Dussourd, and Charles Mitter, "Escalation of plant defense: Do latex and resin canals spur plant diversification?" *American Naturalist* 138, no. 4 (1991): 881–900.

[133] Dietland Müller-Schwarze and Vera Thoss, "Defense on the rocks: Low monoterpenoid levels in plants on pillars without mammalian herbivores," *Journal of Chemical Ecology* 34, no. 11 (2008): 1377.

[134] Yan B. Linhart and John D. Thompson, "Thyme is of the essence: Biochemical polymorphism and multi-species deterrence," *Evolutionary Ecology Research* 1, no. 2 (1999): 151–71.

[135] Daniel Intelmann, Claudia Batram, Christina Kuhn, Gesa Haseleu, Wolfgang Meyerhof, and Thomas Hofmann, "Three TAS2R bitter taste receptors mediate the psychophysical responses to bitter compounds of hops (*Humulus lupulus* L.) and beer," *Chemosensory Perception* 2, no. 3 (2009): 118–32.

[136] Benoist Schaal, Luc Marlier, and Robert Soussignan, "Human foetuses learn odours from their pregnant mother's diet," *Chemical Senses* 25, no. 6 (2000): 729–37.

[137] Sandra Wagner, Sylvie Issanchou, Claire Chabanet, Christine Lange, Benoist Schaal, and Sandrine Monnery-Patris, "Weanling infants prefer the odors of green vegetables, cheese, and fish when their mothers consumed these foods during pregnancy and/or lactation," *Chemical Senses* 44, no. 4 (2019): 257–65.

[138] R. Haller, C. Rummel, S. Henneberg, Udo Pollmer, and Egon P. Köster, "The influence of early experience with vanillin on food preference later in life," *Chemical Senses* 24 (1999):465–67; Delaunay-El Allam, Maryse, Robert Soussignan, Bruno Patris, Luc Marlier, and Benoist Schaal, "Long-lasting memory for an odor acquired at the mother's breast," *Developmental Science* 13 (2010): 849–63.

[139] Martin Jones, "Moving north: Archaeobotanical evidence for plant diet in Middle and Upper Paleolithic Europe," in *The Evolution of Hominin Diets* (Springer, 2009), 171–80.

[140] Joshua J. Tewksbury, Karen M. Reagan, Noelle J. Machnicki, Tomás A. Carlo, David C. Haak, Alejandra Lorena Calderón Peñaloza, and Douglas J. Levey, "Evolutionary ecology of pungency in wild chilies," *Proceedings of the National Academy of Sciences* 105, no. 33 (2008): 11808–11.

[141] Lovet T. Kigigha and Ebubechukwu Onyema, "Antibacterial activity of bitter leaf (*Vernonia amygdalina*) soup on *Staphylococcus aureus* and *Escherichia coli*," *Sky Journal of Microbiology Research* 3, no. 4 (2015): 41–45.

[142] Jean Bottéro, "The culinary tablets at Yale," *Journal of the American Oriental Society* 107, no. 1 (1987): 11–19.

[143] Gojko Barjamovic, Patricia Jurado Gonzalez, Chelsea Graham, Agnete W. Lassen, Nawal Nasrallah, and Pia M. Sörensen, "Food in Ancient Mesopotamia: Cooking the Yale Babylonian Culinary Recipes," in A. Lassen, E. Frahm and K. Wagensonner, eds., *Ancient Mesopotamia Speaks: Highlights from the Yale Babylonian Collection* (Yale Peabody Museum of Natural History, 2019), 108–25.

[144] Won-Jae Song, Hye-Jung Sung, Sung-Youn Kim, Kwang-Pyo Kim, Sangryeol Ryu, and Dong-Hyun Kang, "Inactivation of *Escherichia coli* O157: H7 and *Salmonella typhimurium* in black pepper and red pepper by gamma irradiation," *International Journal of Food Microbiology* 172 (2014): 125–29.

[145] Poul Rozin and Deborah Schiller, "The nature and acquisition of a preference for chili pepper by humans," *Motivation and Emotion* 4, no. 1 (1980): 77–101. The experiment is described in Paul Rozin, Lori Ebert, and Jonathan Schull, "Some like it hot: A temporal analysis of hedonic responses to chili pepper," *Appetite* 3, no. 1 (1982): 13–22.

[146] Paul Rozin and Keith Kennel, "Acquired preferences for piquant foods by chimpanzees," *Appetite* 4, no. 2 (1983): 69–77.

[147] Paul Rozin, Leslie Gruss, and Geoffrey Berk, "Reversal of innate aversions: Attempts to induce a preference for chili peppers in rats," *Journal of Comparative and Physiological Psychology* 93, no. 6 (1979): 1001.

[148] Paul Rozin, "Getting to like the burn of chili pepper: Biological, psychological and cultural perspectives," *Chemical Senses* 2 (1990): 231–69.

[149] Judith R. Ganchrow, Jacob E. Steiner, and Munif Daher, "Neonatal facial expressions in response to different qualities and intensities of gustatory stimuli," *Infant Behavior and Development* 6 (1983): 189–200.

[150] Paul Breslin, "An evolutionary perspective on food and human taste," *Current Biology* 23, no. 9 (2013): R409-418.

[151] Robert J. Braidwood, Jonathan D. Sauer, Hans Helbaek, Paul C. Mangelsdorf, Hugh C. Cutler, Carleton S. Coon, Ralph Linton, Julian Steward, and A. Leo

Oppenheim, "Symposium: Did man once live by beer alone?" *American Anthropologist* 55, no. 4 (1953): 515–26.

[152] Li Liu, Jiajing Wang, Danny Rosenberg, Hao Zhao, György Lengyel, and Dani Nadel, "Fermented beverage and food storage in 13,000 y-old stone mortars at Raqefet Cave, Israel: Investigating Natufian ritual feasting," *Journal of Archaeological Science: Reports* 21 (2018): 783–93.

[153] John Smalley, Michael Blake, Sergio J. Chavez, Warren R. DeBoer, Mary W. Eubanks, Kristen J. Gremillion, M. Anne Katzenberg, et al., "Sweet beginnings: Stalk sugar and the domestication of maize," *Current Anthropology* 44, no. 5 (2003): 675–703.

[154] Katherine R. Amato, Carl J. Yeoman, Angela Kent, Nicoletta Righini, Franck Carbonero, Alejandro Estrada, H. Rex Gaskins, et al., "Habitat degradation impacts black howler monkey (*Alouatta pigra*) gastrointestinal microbiomes," *ISME Journal* 7, no. 7 (2013): 1344–53.

[155] Paulo R. Guimarães Jr., Mauro Galetti, and Pedro Jordano, "Seed dispersal anachronisms: Rethinking the fruits extinct megafauna ate," *PLoS One* 3, no. 3 (2008).

[156] Alcohol is present in most sour fruits. Robert Dudley, "Ethanol, fruit ripening, and the historical origins of human alcoholism in primate frugivory," *Integrative and Comparative Biology* 44, no. 4 (2004): 315–23.

[157] Elisabetta Visalberghi, Dorothy Fragaszy, E. Ottoni, P. Izar, M. Gomes de Oliveira, and Fábio Ramos Dias de Andrade, "Characteristics of hammer stones and anvils used by wild bearded capuchin monkeys (*Cebus libidinosus*) to crack open palm nuts," *American Journal of Physical Anthropology* 132, no. 3 (2007): 426–44.

[158] Matthias Laska, "Gustatory responsiveness to food-associated sugars and acids in pigtail macaques, *Macaca nemestrina*," *Physiology and Behavior* 70, no. 5 (2000): 495–504.

[159] D. Glaser and G. Hobi, "Taste responses in primates to citric and acetic acid," *International Journal of Primatology* 6, no. 4 (1985): 395–98.

[160] Daniel H. Janzen, "Why fruits rot, seeds mold, and meat spoils," *American Naturalist* 111, no. 980 (1977): 691–713.

[161] Matthew A. Carrigan, Oleg Uryasev, Carole B. Frye, Blair L. Eckman, Candace R. Myers, Thomas D. Hurley, and Steven A. Benner, "Hominids adapted to metabolize ethanol long before human-directed fermentation," *Proceedings of the National Academy of Sciences* 112, no. 2 (2015): 458–63.

[162] Rotten fruits might also have contained insects and the additional protein provided by their bodies. Some primates actually appear to prefer fruits that contain insects to those that don't. Kent H. Redford, Gustavo A. Bouchardet da Fonseca, and Thomas E. Lacher, "The relationship between frugivory and insectivory in primates," *Primates* 25, no. 4 (1984): 433–40.

[163] A. N. Rhodes, J. W. Urbance, H. Youga, H. Corlew-Newman, C. A. Reddy, M. J. Klug, J. M. Tiedje, and D. C. Fisher, "Identification of bacterial isolates obtained from intestinal contents associated with 12,000-year-old mastodon remains," *Applied Environmental Microbiology* 64, no. 2 (1998): 651–58.

[164] Elizabeth Wason, "The Dead Elephant in the Room" *LSA Magazine* (2014) https://lsa.umich.edu/lsa/news-events/all-news/search-news/the-dead -elephant-in-the-room.html.

[165] Iwao Ohishi, Genji Sakaguchi, Hans Riemann, Darrel Behymer, and Bengt Hurvell, "Antibodies to *Clostridium botulinum* toxins in free-living birds and mammals," *Journal of Wildlife Diseases* 15, no. 1 (1979): 3.

[166] Daniel T. Blumstein, Tiana N. Rangchi, Tiandra Briggs, Fabrine Souza De Andrade, and Barbara Natterson-Horowitz, "A systematic review of carrion eaters' adaptations to avoid sickness," *Journal of Wildlife Diseases* 53, no. 3 (2017): 577–81.

[167] Daniel C. Fisher, "Experiments on subaqueous meat caching," *Current Research in the Pleistocene* 12 (1995): 77–80.

[168] John D. Speth, "Putrid meat and fish in the Eurasian Middle and Upper Paleolithic: Are we missing a key part of Neanderthal and modern human diet?" *PaleoAnthropology* 2017 (2017): 44–72.

[169] William Sitwell, *A History of Food in 100 Recipes* (Little, Brown, 2013).

[170] Mark Kurlansky, *Salt* (Random House, 2011).

[171] Adam Boethius, "Something rotten in Scandinavia: The world's earliest evidence of fermentation," *Journal of Archaeological Science* 66 (2016): 169–80.

[172] Sveta Yamin-Pasternak, Andrew Kliskey, Lilian Alessa, Igor Pasternak, Peter Schweitzer, Gary K. Beauchamp, Melissa L. Caldwell, et al., "The rotten renaissance in the Bering Strait: Loving, loathing, and washing the smell of foods with a (re)acquired taste," *Current Anthropology* 55, no. 5 (2014): 619–46.

[173] Hsiang Ju Lin and Tsuifeng Lin, *The Art of Chinese Cuisine* (Tuttle Publishing, 1996).

[174] Cristina Izquierdo, José C. Gómez-Tamayo, Jean-Christophe Nebel, Leonardo Pardo, and Angel Gonzalez, "Identifying human diamine sensors for death related putrescine and cadaverine molecules," *PLoS Computational Biology* 14, no. 1 (2018): e1005945.

[175] Paul Kindstedt, *Cheese and Culture: A History of Cheese and Its Place in Western Civilization* (Chelsea Green Publishing, 2012).

[176] Benjamin E. Wolfe, Julie E. Button, Marcela Santarelli, and Rachel J. Dutton, "Cheese rind communities provide tractable systems for in situ and in vitro studies of microbial diversity," *Cell* 158, no. 2 (2014): 422–33.

[177] David Asher, *The Art of Natural Cheesemaking* (Chelsea Green Publishing, 2015).

[178] Gordon M. Shepherd, *Neuroenology: How the Brain Creates the Taste of Wine* (Columbia University Press, 2016).

[179] David G. Laing and G. W. Francis, "The capacity of humans to identify odors in mixtures," *Physiology and Behavior* 46, no. 5 (1989): 809–14.

[180] Masaaki Yasuda, "Fermented tofu, tofuyo," in *Soybean—Biochemistry, Chemistry and Physiology*, ed. T. B. Ng (InTech, 2011), 299–319.

[181] From an email on February 8, 2020.

[182] Roman M. Wittig, Catherine Crockford, Tobias Deschner, Kevin E. Langergraber, Toni E. Ziegler, and Klaus Zuberbühler, "Food sharing is linked to urinary oxytocin levels and bonding in related and unrelated wild chimpanzees," *Proceedings of the Royal Society B: Biological Sciences* 281, no. 1778 (2014): 20133096.

[183] Ammie K. Kalan and Christophe Boesch, "Audience effects in chimpanzee food calls and their potential for recruiting others," *Behavioral Ecology and Sociobiology* 69, no. 10 (2015): 1701–12.

[184] Ammie K. Kalan, Roger Mundry, and Christophe Boesch, "Wild chimpanzees modify food call structure with respect to tree size for a particular fruit species," *Animal Behaviour* 101 (2015): 1–9.

[185] Martin Jones, *Feast: Why Humans Share Food* (Oxford University Press, 2007).

ILLUSTRATION CREDITS

Figure P.1: The authors.

Figure 1.1: Data from an unpublished manuscript led by Mick Demi.

Figure 1.2: Photo by Wei Fuwen.

Figure 2.1: Photo by Liran Samuni as part of the Taï Chimpanzee Project.

Figure 2.2: Figure based on a similar figure in Robert R. Dunn, Katherine R. Amato, Elizabeth A. Archie, Mimi Arandjelovic, Alyssa N. Crittenden, and Lauren M. Nichols. "The internal, external and extended microbiomes of hominins," *Frontiers in Ecology and Evolution* 8 (2020): 25. Original data from Hjalmar S. Kühl, Christophe Boesch, Lars Kulik, Fabian Haas, Mimi Arandjelovic, Paula Dieguez, Gaëlle Bocksberger et al., "Human impact erodes chimpanzee behavioral diversity," *Science* 363, no. 6434 (2019): 1453–55.

Figure 2.3: Photo by Liran Samuni.

Figure 2.4: Photo by Alex Wild.

Figure 3.1: Photo by Daniel Mietchen, Creative Commons. https://commons .wikimedia.org/wiki/File:Iowa_Pig_(7341687640).jpg.

Figure 4.1: Photo taken by the archaeologist Gary Haynes.

Figure 4.3: Photo by Menuka Scetbon-Didi.

Figure 4.4: Data are from Jeremy M. Koster, Jennie J. Hodgen, Maria D. Venegas, and Toni J. Copeland, "Is meat flavor a factor in hunters' prey choice decisions?" *Human Nature* 21, no. 3 (2010): 219–42.

Figure 4.5: Data for the Waorani are from Sarah Papworth, E. J. Milner-Gulland, and Katie Slocombe, "The natural place to begin: The ethnoprimatology of the Waorani." *American Journal of Primatology* 75, no. 11 (2013): 1117–28.

Figure 5.1: Creative Commons photo.

Figure 5.2: Robert J. Warren, "Ghosts of cultivation past-Native American dispersal legacy persists in tree distribution," *PloS One* 11, no. 3 (2016).

Figure 6.1: Photograph by Martin Oeggerli.

Figure 6.2: Data are from Paul W. Sherman and Jennifer Billing, "Darwinian gastronomy: Why we use spices: Spices taste good because they are good for us," *BioScience* 49, no. 6 (1999): 453–63.

Figure 6.3: Data are from Paul W. Sherman and Jennifer Billing, "Darwinian gastronomy: Why we use spices: spices taste good because they are good for us," *BioScience* 49, no. 6 (1999): 453–63.

Figure 7.1: Photo by Liz Rashee.

Figure 8.1: Photo by Jose Bruno-Bárcena.

Figure 8.3: Data from Maria Dembińska, "Diet: A comparison of food consumption between some eastern and western monasteries in the 4th–12th centuries," *Byzantion* 55, no. 2 (1985): 431–62.

Figure 9.1: Photo by Liran Samuni, Taï Chimpanzee Project.

INDEX

A page number in *italics* refers to a figure or table.

absinthe, 16
acetic acid (vinegar): aroma of fermented herring and, 179; sour taste of, 164
acetic acid bacteria, 156, 162, 165, 166
acidic foods, free of pathogens, 162, 166
adrostenol, 58, 59
adrostenone, 229n21
agouti meat, 102
alcohol: favoring some bacteria over others, 166, 202; in fermented foods and drinks, 156–57; intoxication by, 166–67; metabolism in liver of hominids and, 166–67; pleasure induced by, 167, 169, 180
alcohol dehydrogenase, 166, 180
algae, eaten by chimpanzees, *30*, 34
allicin, 141–42
alliin, 141, 234n8
alliinase, 141, 235n8
alliums, 141–43. *See also* garlic
almendro fruits, 159–60, *161*
Amato, Katie, 35–36, 156, 158–59, 160–62, 165–66, 167
America's First Cuisines (Coe), 105
amino acids: nitrogen in, 9, 10, 11; produced by bacteria, 35; tasted by hummingbirds, 21

ammonia aroma, of fermented shark, 239n12
amylase, 14
ancient humans: atrophied food-processing body parts of, 49–50; defined, 32; with different diets in different regions, 50–51; experience of flavor in, 69; fermentation of fruits by, 167–69, 180; fueling a larger brain with new food ways, 32–33, 36; honeybees and, 32–33; hunting animals of Europe and Asia, 83; learning new aromas and flavors, 75–76, 77–78; processing foods, 32, 34–37; shellfish in diets of, 33–34, 48, 51; with smaller teeth than chimpanzees, 32, 33, 49–50; tools of, 34–35, 77–78. *See also* *Homo erectus*
Ancient Mesopotamia Speaks (Lassen, Frahm, and Wagensonner), 142
anise, 130; eaten in pregnancy, 135, 140
ants: eaten by primates, 41, 43; fruits appealing to, 116
Apicius, spiced meat recipe of, 144–45
apple cider, 168
Arandjelovic, Mimi, 164, 223n25

armadillos: bear-sized (glyptodonts), 122, 123; meat of, 102

aromas: of alcoholic fruits, 169; of animal's food appearing in their meat, 99, 102; of cheeses, 193, 197, 199–200, 240n3; complex, 48, 70–72, 168, 199–201; as component of flavor, 6, 53–54, 67; differences among species in perception of, 54; evolution of sense of smell and, 63–64, 66–67; of fermented foods, 167, 168, 172, 173, 179–80; of fish, 136; of fruits attractive to mammals, 116; of greenness, 136; hardwired in mammalian brains, 57–58; individualized categorization of, 74–75, 226nn13–14; innate human predispositions toward, 69–70; learning to identify, 73, 74, 199–200; loss of transverse lamina and, 66–67; path of exhalation in humans and, 68; prenatal and neonatal experiences of, 134–37, 139–40, 233nn4–5, 234n6; ranked in memory, 75; of toxic plants, 133–34; of truffles, 62–63. *See also* olfactory bulb; olfactory receptors; orthonasal aromas; retronasal aromas

The Art of Natural Cheesemaking (Asher), 196

Asher, David, 196

Asiago, 192

aspirin, 16

Australopithecus: bipedalism changing olfaction in, 67–68; evolutionary changes in, 31; experience of flavor in, 69; forest diet of *A. sediba*, 219n7; honeybees and, 33

avocados: enjoyed by cats, 96; mouthfeel of fats in, 15, 96;

undispersed seeds in wild relatives of, 118–19

awamori, 201

Axel, Richard, 73

Aynaud, Carole, 60

Aynaud, Edouard, 60–61, 62

bacteria, fermenting. *See* acetic acid bacteria; lactic acid bacteria

bacteria, pathogenic: killed by acid, 162, 166; killed by alcohol, 166

Baker, Samuel White, 109

Barkai, Ran, 108–9, 110, 184, 230n28

barley beer, 157, 158, 236n2

Bates, Henry, 102

bats, aromas of fruits appealing to, 116

bears: meat of, 229n17; in megafauna, 88

beer: brewing process for, 236n2; cheeses washed in, 199, 201; preceding onset of agriculture, 157–58; sour, 156, 157, 162. *See also* hops

bees. *See* honeybees

Benedictine monks: cheeses made by, 189–90, 197; diet of, 189, *195*, 201, 241n11; rule book for, 188, *191*, 241n4. *See also* monasteries

Berbesque, Colette, 44–46, 229n19

big mammals: in cave paintings, 60, 112–13; Clovis people's killing and eating of, 84–88; encountered by first Americans, 83–84. *See also* megafaunal extinctions

Billing, Jennifer, 143

biological stoichiometry, 3–5; salt and, 7

bipedalism: energy saving derived from, 218n3; evolution of, 30, 31; olfaction changed by, 67–68

birds: able to taste sugars, 21, 22; attracted to colorful fruits, 116;

capsaicin in chilies and, 149, 235n13; eaten into extinction, 89; fat in some seabirds, 229n15; meat of, 101, 229n18
A Bite-Sized History of France (Hénaut and Mitchell), 199
bitter leaf, as medicine and spice, 139, 234n7
bitter taste: added to dangerous products, 217n13; of hops in beer, 17, 133, 146; of leaves eaten by howler monkeys, 102–3; newborn baby's response to, 234n5; not detected by some monkeys, 22; reduced by fermentation, 168; sometimes ignored for coffee or beer, 17; stronger in children, 217n14; of toxic seeds spit out by monkeys, 231n2; of toxins intentionally used, 134
bitter taste receptors, 15–17; differing among animal species, 96, 133; differing between humans and chimpanzees, 39, 220n14; differing between individual people, 16; differing between recent human lineages, 51, 224n27; nausea and vomiting associated with, 17; phenylthiocarbamide and, 224n27; toxic plants and, 133–34
black pepper, 131, 147–48, 235n13
Blake, Michael, 158
bloomy-rind cheeses, 194, *198*
blue cheeses, 195–96. *See also* Cabrales cheese
Boesch, Christophe, 27, 164
Boethius, Adam, 177–78
bones: DNA extracted from, 51, 223n26; eating marrow and grease of, 51
bone tools, of first people in the Americas, 84

Booth, Alvin, 203, 204, 208
Boswell, James, 25, 80
Bracciolini, Poggio, 2–3
brain: aromas that are hardwired in, 57–58; cataloging aromas, 72, 73–75, 78, 225n10, 226n11; evolving larger size in ancient humans, 31, 32, 33, 36, 50; receiving signals from taste receptors, 6–7. *See also* learning; olfactory bulb
Breslin, Paul, 155
Brevibacterium linens, 196, 197, *198*
Brie de Meaux, 194, 202
Brillat-Savarin, Jean Anthelme, xi–xii; on cooking, 220n11; on distinguishing flavors, 80; on eating oysters, 34, 47–48; on eating slowly, 223n24; on foods disliked by French, 223n22; on happiness from eating new dish, 83; on human experience of flavor, 68–69; on infinite number of flavors, 129; on needed nutrients, 4; on sense of smell, 53; on sociability of meals, 209–10; on things of no importance, 41; on uses of taste, 1
browning of cooked food, 71, 225n7
Bruno-Bárcena, Jose, 182–83, 186, 188
Buck, Linda, 73
buffalo milk cheese, 191
bulbs: defined, 219n4; giving flavor to peccary meat, 101; as spices, 131
Bunyard, Edward, 182, 193, 240n1
butterflies and moths, sex pheromone of, 225n3
Byron, Lord, 114

Cabrales cheese, 182–88, *187*, 195–96, 202
cadaverine, 57, 58, 225n2
caffeine, 133

calcium, possible taste receptor for, 13
Camembert, 135, 194
capsaicin, 148, 149
capuchin monkeys, 159–60, *161*, 168
carbohydrates: broken down in
    rumen, 100; complex, 14; as energy
    source, 14, 98
carbon compounds, metabolized for
    energy, 14
carnivores: broken sweet taste receptors
    of, 18–19; storing and fermenting
    meat, 238n9; taboo against eating
    meat of, 228n10
carrion-feeding animals, 58, 106, 163,
    171, 228n10
Catching Fire (Wrangham), 36
Cato the Elder, on salting ham,
    176–77
cats: broken bitter taste receptors of,
    96; broken sweet taste receptors of,
    18–19, 96; enjoying avocados, 96
cave paintings: of Cantabria, 183–84;
    of the Dordogne, 60, 111–13
Chapman, Ben, 147
Cheese and Culture (Kindstedt), 199
cheeses: aromas of, 193, 197, 199–200,
    240n3; easier to make hard than
    soft, 186; fresh, 190–91; influenced
    by animal's diet, 190–91, 240n3;
    learning to discriminate between,
    200; made by monasteries, 189–90,
    193–94, 196–201, 241n8; as means
    for storing milk, 186; shipping
    and storing, 186, 191; sulfurous,
    eaten during pregnancy, 135–36.
    See also hard cheeses, aged; soft
    cheeses, aged
chemesthesis, 147–52
chemical elements: in animals vs.
    plants, 5, 8–10, *9*; in mammals, 4–5

chewing: to experience full flavors,
    223n24; food softened by fermen-
    tation and, 46–47, 168, 180; made
    easier by processing, 47; of shellfish,
    47–48
chili peppers, 149–52, 235nn12–13,
    236n14; attracting birds, 149, 235n13;
    as fruits, 130–31
Chimay cheese, 199, 201
chimpanzees: alcohol metabolism in,
    166–67; calls of, 207–8; crabs eaten
    by, 34; culinary traditions of, 28–29,
    37, 218n1; eating colobus monkeys,
    90, 205, *209*, 242n4; eating figs, *44*,
    221n15; enjoying sour taste, 38, 164–65;
    experience of flavor in, 67; as gastro-
    nomes, 41, 43; Goodall's studies of,
    25–27; human body's differences
    from, 37–38; human researchers
    eating foods of, 39–40; learning to
    enjoy new fruits, 76–77; meat in diet
    of, 48–49, 222n20, 223n25 (*see also*
    colobus monkeys); not mixing
    multiple ingredients, 130; oxytocin
    in, 206, 243n5; plants eaten by, 39–40,
    220nn13–14, 221n15; preferring
    cooked vegetables and meat, 48–49,
    223n25; prenatal learning of benefi-
    cial foods and, 136; seeking flavors,
    not necessarily nutrients, 41, 42–43,
    222n18; sharing food, 205–6, *209*,
    242n4; sour taste and, 38, 164–65;
    taste receptors similar to human
    ones, 23, 37–38; tools of, xiv, 26–29,
    29, *30*, 32, 35, 37, 41; using plants as
    medicine, 234n7. *See also* common
    ancestor of humans and chimpanzees
Chinese Gastronomy (Lin and Lin),
    46, 70
cider, 168

cinnamon, 148, 235n12
citric acid, *18*, 164
climate change, and megafaunal
    extinctions, 90, 228nn7–8
Clostridium botulinum, 171
Clostridium perfringens, 171
Clovis people, 82; changed after
    megafaunal extinctions, 90; eating
    large quantities of meat, 84, 86–88;
    Fisher's speculations on fermenta-
    tion by, 169–75; hunting megafauna,
    84–88, 91, 119; killing predators not
    eaten, 95; likely food preferences
    of, 105–8; limits of our information
    about, 103–5; pursuing pleasures of
    food, 110–11
Clovis points, 82, 84, *85*, 90
Coe, Sophie, 105
collagen, in meat, 98
colobus monkeys, eaten by chimpan-
    zees, 90, 205, *209*, 242n4
common ancestor of humans and
    chimpanzees: enjoying honey, 44; as
    gastronomes, 41, 43; prenatal learn-
    ing of beneficial foods and, 136;
    seeking flavorful foods, 40–41; using
    newer kinds of tools, 43–44, 46–47;
    using plants as medicine, 234n7
complex aromas, 48, 70–72, 168, 199–201
complex carbohydrates, 14
complex flavors, 200–201
cooked meat: aromas of, 48, 70–72;
    Brillat-Savarin on, 220n11; of Clovis
    people, 82, 86, 87, 104–5; compared
    to raw meat, 47–48; of Fisher's fer-
    mented horse, 173; in Homer's *Iliad*,
    86–87; increased glutamate in, 48;
    Neanderthals and, 78–79, 87; from
    older animals, 104–5; of peoples de-
    scended from Clovis, 87, 227n4;

preferred to raw by chimpanzees,
    48–49, 223n25; sources of flavor in,
    97–98; spices used with, 143–45, *144*.
    See also stews
cooking: earliest history of, 86; at high
    temperatures, 71; improving flavor,
    36–37, 48–49, 72; Maillard reaction
    and, 71, 177; reducing chewing time,
    46–47. *See also* cooked meat; fire
cooking pots, 139
corn: fermented by ancient Native
    Americans, 158; spices used with,
    147, 153
Craig, Oliver, 137–39
Crittenden, Alyssa, 220n12
Croatia, stone pen in, ix–x, *xviii*
Cro-Magnon humans, 111
cuisine: of chimpanzees, xiv, 28–29, 37;
    defined, 28
culinary endangerment, 103
culinary extinctions, 79, 89–90
culinary traditions: of chimpanzees,
    28–29, 37, 218n1; defined, 28
culture and diet, 28
cutting foods, 46–47, 78

dandelions, 134
Darwin, Charles, 25, 26, 102, 210–11
dashi, 10–11, 217n9
deer meat, 100
deliciousness, xii; beginning of cooking
    and, 36–37; of cheese that's hard to
    make, 186; improved by together-
    ness, 210; of proboscideans, 108–10;
    tool use by our ancestors and, 23–24.
    See also flavor; pleasure
Denisovans, 51
De rerum natura. See Lucretius
The Descent of Man (Darwin), 25
desert mammals, eating salty plants, 22

dill, 134

dimethyl sulfide: babies preferring
aroma of, 136; from truffles, 59

Dinner with Darwin (Silvertown), 11

displeasure: bitter taste receptors and,
15–17; excess of salt and, 8; Lucretius
on, 1–2, 3; of medium-length fatty
acids, 217n12; memory of an aroma
and, 75

disulfide bonds, aroma associated
with, 225n4

dodo, eaten into extinction, 89

dogs: ancient relationship to kitchen
and, 225n5; chemicals deposited in
fat eaten by, 99; human experience
of flavor compared to, 69; hunting
truffles, 54, 59, 60–62, 65, 69; noses
of, 64–65

dolphins, broken taste receptors of, 19

dopamine: sensation of pleasure and,
6; social bonds and, 206

Dordogne: cave art in, 60, 111–13;
hunting truffles in, 60–62; Neanderthals in, 59–60, 78–79, 111, 231n31

drying meat, 175–76

duiker meat, 102

Efe hunter-gatherers, 33

elephant meat, 108–10; of delicious
feet, 109–10, 113; fermented in
ground for 17 years, 238n8

elephants: hunted with long bows,
231n29; salt receptors of, 7; sex
pheromone of, 58, 225n3

endorphins: pain of eating chili peppers
and, 150; signals from taste receptors
and, 6

energy: from carbohydrates, 14, 98;
from fats, 14, 98; processing of food
and, 34, 35, 36

Entangled Life (Sheldrake), 56

Epicurus, 2

Époisses, 135, 198–99

Ertebølle hunter-gatherers, 138–39

Estienne, Vittoria, 222n18

ethanol. *See* alcohol

Evans, Josh, 230n22

extinct animals. *See* megafaunal
extinctions

extinct flavors, 88

extinct plants, 227n6

fat: chemicals held in, 98, 99, 100;
cultural differences in liking of,
105; energy from, 14, 98; factors
affecting an animal's amount of, 98,
228n14; fermented, 105; of fruits
attractive to mammals or ants, 116;
in meat, 98–99, 105, 107; in meat of
elephant feet, 109, 110; in meat of
some seabirds, 229n15; mouthfeel of,
15, 98, 105; pleasurable to mammals,
14–15; as triglycerides, 217n12

fatty acids: adding flavor to meat of
ruminants, 100; in goat cheese and
buffalo cheese, 191; tastes of, 98,
217n12

fear, aromas hardwired for, 57–58

Feast (Jones), 211

fermentation: acidic foods and drinks
made by, 156–57; acidity as indicator
of safety and, 161–62; before agriculture, 158; alcoholic foods and
drinks made by, 156–57; aromas of
foods and, 167, 168, 172, 173, 179–80;
benefits of, 35–36, 168; of fish, 177–81;
of fish flakes (katsuobushi), 10–11,
216n9; of fruits, 156, 158, 160–62,
167–69, 180, 237n6; improving
flavor, 168; making food soft for

chewing, 46–47, 168, 180; of meat, 35, 48, 169–75, 177–81, 238n9; microbiological definition of, 156; nutrition enhanced by, 168; in ruminants' digestive tract, 100; for storing fruits and vegetables, 168; of tofu, 201–2. *See also* beer; cheeses

fire, 36–37, 78, 220nn11–12; communication at gatherings around, 210. *See also* cooking

first people in the Americas, 82–83

fish: eaten while breast-feeding, 136; fermentation of, 177–81, 239nn11–12

Fisher, Daniel, 169–75, 177

fish flakes (katsuobushi), 10–11, 216n9

fish sauces, 179, 239n13

Fjeldså, Jon, 100, 101, 102, 229n15, 229n18

flavor: of aged soft cheeses, 193; aromas as component of, 6, 53–54, 67; complex, 200–201; components of, 6; food preferences of chimpanzees and, 40–41; guiding animals to their needs, 5–6; improved by cooking, 36–37, 48–49, 72; loss of transverse lamina and, 67; more available with processing, 34; prenatal and neonatal experiences of, 134–37, 139–40, 233nn4–5, 234n6. *See also* deliciousness

flavor of meat: bearing flavors of animal's diet, 100–103, 229n18; eaten by hunter-gatherers, 91–97, 97, 113; fat and, 99, 228n14; fermentation and, 168; fruits eaten by animal and, 100, 101, 102, 106, 113; of herbivores, 99–103; muscle and, 97–98, 106–7; of omnivores, 99–103, 229n17; of predators, 99; of ruminants, 100, 102, 105; sources of, 97–98; of white

meat vs. red meat, 106–7. *See also* cooked meat

flavor-seeker hypothesis, 37

food-borne illness: antimicrobial compounds in spices and, 140–41, *141*, 143, 145, 146, 153; aromas associated with, 140; black pepper as source of, 147; studied by specialists, 212

Frank, Hannah, 163, 164

fresh cheeses, 190–91

fructose, 14; in honey, 45

fruits: complex aromas of, 72; evolved for dispersion of seeds, 115–17; fats in, 116; fermentation of, 156, 158, 160–62, 167–69, 180, 237n6; giving flavor to meat, 100, 101, 102, 106, 113; qualities appealing to different animal groups, 116; spices in form of, 130–31 (*see also* chili peppers); undispersed, 118–19. *See also* megafauna fruits

fungi. See *Penicillium* fungi; truffles

Garcia effect, 140

garlic, 134, 141–43, 234n8; in amniotic fluid, 135, 233n4; antimicrobial properties of, 141, 142, 145–46

garlic mustard, 138–39

garum, 239n13

gastronomy, xi–xiii, 41

gastrophagy, 107

Gastrophysics (Spence), xiii

giant mammals. See big mammals; megafaunal extinctions

giant pandas, 19–20, *20*, 218n16

giant sloths: of Costa Rica, 121, 123; killed by Clovis people, 84, 88; questionable flavor of, 230n24. *See also* sloths

glucose, 14; in honey, 45
glutamate: freed by cooking or
    fermenting meat, 48; taste
    threshold of, 18
glutamic acid, 11, 12; formed in
    fermentation, 168
glyptodonts, 122, 123
goat milk and cheese, 191
gomphotheres, 88, 106, 108, 122, 123
Goodall, Jane, 25–27
gorillas: alcohol metabolism in,
    166–67; experience of flavor in, 67;
    food preferences of, 38, 221nn15–17;
    mutation in sweet taste receptors,
    41–42, 221n17; preferring cooked
    vegetables, 48–49; sour taste en-
    joyed by, 165; using plants as
    medicine, 234n7
Gotelli, Nick, 211
Gouda, 192, 241n6
grains: domesticated in order to
    ferment, 158; spices to flavor simple
    dishes of, 147, 153
grapes, fermentation of, 237n6
grasses: giving flavor to cheeses,
    190–91; in mammoth diet, 106, 108
grassland plants, silica in, 131
grinding, 46–47, 78
grizzly bear meat, 229n17
grouse, 101, 102
Gruyère monastery, 198
guanylate, 11, 12
Guénard, Benoit, 201
Guevara, Elaine, 42, 221n17
Guthrie, Dale, 230n23

hackberries, 137
Hadza hunter-gatherers, 44, 101,
    220n12, 223n22, 229n19
Halwachs, Winnie, 117

ham, salting of, 176–77
haplorhine primates, evolutionary
    changes in, 65–69
hard cheeses, aged, 190, 191–92, 199
Harrison, Jim, 80–81, 227n1
Haynes, Gary, 84, 109, 229n17, 230n28
Haynes, Vance, 82
heat receptors: of birds or rodents,
    149; in mouth and nose, 148
Hénaut, Stéphane, 199
Henry, Amanda, 219n7
herbivores: balance of chemical
    elements in, 5, 9–10; chemical
    defenses of plants and, 131–34;
    flavor of meat of, 99–103; phospho-
    rus and, 12; seeking out salt, 7
herbs, 130; meat of animals feeding on,
    101
herring, fermented, 178–79, 180,
    239n11
Holmberg, Allen, 130
Homer's *Iliad*, sacrifice of cattle in,
    86–87
Homo erectus, 31–32; fermentation and,
    167; olfactory libraries of, 75–76;
    protein from fossil teeth of,
    223n26
Homo sapiens, 51, 213
honeybees: calmed by smoke or
    plant exudates, 32–33, 44, 220n9;
    chimpanzees accessing honey of,
    32, 42–43, 222n18; chimpanzees
    eating bee brood, 42, 222n18; Hadza
    hunter-gatherers and, 44–46; honey-
    making process of, 45
honey locust trees, 126–27, 127
hops, 17, 133, 146
horse meat: fermented in Fisher's
    experiment, 172–75; flavor of,
    100–101

horseradish, 148
horses: in cave paintings, 60, 112; giant, 88; Janzen's fruit experiments with, 122–24; preferring sweet to sour or salty taste, 232n5
howler monkeys, disliked meat of, 95, 96, 102–3
human ancestors, 30–31. *See also* ancient humans; common ancestor of humans and chimpanzees; recent humans
hummingbirds, 21
hunter-gatherers: Efe people, 33; Ertebølle people, 138–39; fires of, 220n12; first people in the Americas, 83; flavors of meat eaten by, 91–97, 97, 113; Hadza people, 44, 101, 220n12, 223n22, 229n19; Mayangna and Miskito in Nicaragua, 93–97, 94, 97; pounding food, 35; spices and, 130, 137–39. *See also* Clovis people
hunters: choosing prey with preferred flavor, 95, 101; of Europe and Asia for a million years, 83; first people in the Americas, 83–84; optimal foraging and, 91, 93, 94, 95. *See also* megafaunal extinctions
Hutson, Jarod, 84

Ikeda, Kikunae, 10–12
inosinate, 11, 12
insects as food: ants eaten by primates, 41, 43; concentrating flavors of their diet, 230n22

jamón ibérico, 177
Jänig, Susann, 67
Janzen, Daniel, 116–24, 232n6
Japanese monks, 201

jicaro fruits, 122–24
Jones, Martin, 211

Kalan, Ammie, 207
katsuobushi, 10–11, 216n9
Kays, Roland, 230n24
KCNK receptors, 148–49
Kindstedt, Paul, 193
Koko the gorilla, 49
kombucha, 157, 160, 161
Koster, Jeremy, 92–96, 98–99, 228nn10–11, 228n14
Kuehl, Hjalmar, 46, 223n25
Kurlansky, Mark, 177

lacrimators, 142
lactase, 38
lactic acid: consumed by fungi on cheese, 196, 241n9; sour taste of, 164
lactic acid bacteria, 156, 162, 165, 166; cheeses and, 196, 241n9; in Fisher's fermenting meat, 173, 174, 175
Lactobacillus: in Fisher's fermenting meat, 174, 175; fruits made sour by, 167
Lambert, Joanna, 106
Lanning, Nike, 204, 208
learning: chimpanzees enjoying new fruits and, 76–77; by Clovis people, 86; of cooking methods by our ancestors, 227n3; to enjoy repulsive aromas, 180; to enjoy spicy food, 151, 152; to identify aromas, 73, 74, 199–200; to like hops in beer, 146; to like or dislike aromas, 139–40, 145; to love flavors, 72; of new aromas and flavors by ancient humans, 75–76, 77–78; of reaction to sour taste, 155

leaves: spheres of chemicals on, 130, 132, 233n2; of trees, and flavor of meat, 102–3, 106

leftovers, plant parts used for preservation of, 139

lemons, 77, 131, 162, 164

Lévi-Strauss, Claude, 227n4

Li, Xia, 19

Lieberman, Daniel, 39, 66, 68

Lin, Hsiang Ju, 46, 70

Lin, Tsuifeng, 46, 70

Liu, Li, 157

Lost Feast (Newman), 89

Lucretius, 1–3, 4; on atoms of foods, 8; bitter tastes and, 16; on different senses in different creatures, 17; on odors, 53; on a swerve, 23

Madden, Anne, xvii

Maillard, Louis Camille, 71

Maillard reaction, 71, 177

Mallot, Liz, 159–60, *161*, 237n3

mammals: chemical elements in, 4–5; fruit qualities with attraction for, 116. *See also* big mammals

mammoth meat, 80, 230n28; of delicious feet, xv, 110, 129–30; imagined cuts of, *92*; of red muscle, 107

mammoths: in cave paintings, 60, 112; climate change and, 228nn7–8; killed by Clovis people, 84, 88, 90, 106, 110, 238n7; stone tools found with bones of, 82; surviving until 2000 BCE, 89; woolly, 89, 108, 228n8

Manchego, 192

manna of the Bible, 10

Maroilles, 198

Martin, Paul S., 88–89, 95, 119, 121–22

Martinez del Rio, Carlos, 105

mastodons: apparently stored by Clovis people, 169–70, 174–75; with bone point in rib, 84; climate change and, 228n7; delicious meat of, 108; fruit-eating, 121–22; killed by Clovis people, 84, 88, 106, 169–70, 238n7

Mattes, Richard D., 217n12

Maupassant, Guy de, 100–101

Mauritius red rail, eaten into extinction, 89

Mayangna people, 92–93; food preferences of Waorani and, 103, *104*; tastiness of different animals and, 97, 100, 101–3

McGee, Harold, 46, 70, 72

meat: with aromas from animal's food, 99, 102; in chimpanzee diet, 48–49, 222n20, 223n25 (*See also* colobus monkeys); connective tissue in, 33; cut to facilitate digestion, 35; dangerous bacteria in decay of, 171; eaten by Clovis people, 84, 86–88; eaten by Neanderthals, 78–79, 84, 86; fat in, 98–99, 105, 107, 109, 110, 229n15; fermentation of, 35, 48, 169–75, 177–81, 238n9; with flavors from animal's food, 102–3, 229n17; of organs, 107; processing of, 47; raw, 34, 46, 223n23; red vs. white, 106–7. *See also* cooked meat; flavor of meat

megafauna fruits, 113, 114, 118–24, 232n6; of almendro tree, 159–60; dispersed after megafaunal extinction, 124–28; human contribution to survival of, 125–28; stinking toe tree, 118–20, *120*, 128

megafaunal extinctions, xv; causes of, 90–91; climate change and, 90,

228nn7–8; as culinary extinctions, 89; ecosystem changes caused by, 90; in Europe, 89, 112–13; hunting by Clovis people and, 88, 90–91, 119; of non-ruminants vs. ruminants, 230n23; undispersed fruits and, 119–26. *See also* big mammals

menthol, 72, 73, 148

milk: flavors from cooking at high temperatures and, 225n8; lactase in adults and, 38; microbes from udders and skin contained in, 240n3; plant compounds contained in, 240n3; stored by making cheese, 186

mint, 73, 75, *132*, 134, 148

Miskito people, 92–93, *94*; food preferences of Waorani and, 103, *104*; tastiness of different animals and, *97*, 100, 101–3

Mitchell, Jeni, 199

moas, eaten into extinction, 89

monasteries: cheeses made by, 189–90, 193–94, 196–201, 241n8; origin of, 188. *See also* Benedictine monks

monkey meat: of colobus eaten by chimpanzees, 90, 205, *209*, 242n4; of disliked howler monkeys, 95, 96, 102–3; of fruit eating species, 100, *102*, 103

monkeys and sour tastes, 163–64

Mouritsen, Ole, xiii

mouthfeel: of aged soft cheeses, 193; as component of flavor, 6, 46, 68; defined, 15; diverse experiences of, 46; of fat, 15, 96, 98, 105; improved by processing, 46, 47; of muscle in cooked meat, 97

Mouthfeel (Mouritsen and Styrbæk), xiii

MSG, 11–12

Munster, 135, 198, 241n12

muscle, flavor of, 97–98, 106–7

mustard, 148

Nabhan, Gary, 81, 101

Navajo, eating meat of animals feeding on sage, 101

Neanderthals: coexisting with *Homo sapiens*, 231n31; cooking meat, 78–79; diet in a Gibraltar cave, 226n15; in the Dordogne, 59–60, 78–79, 111, 231n31; eating meat, 84, 86; elephants butchered by, 109–10; experience of flavor in, 69; hackberries found on hearth of, 137; hunting animals of Europe, xv, 83; taste receptors of, 51; tasting of phenylthiocarbamide and, 224n27

Neuroenology (Shepherd), xiii, 75, 199–200

Neurogastronomy (Shepherd), xiii, 64

Newman, Lenore, 89

night monkeys, liking sour taste, 164

Nishida, Toshisada, 39–40, 76, 164–65, 220n14, 221n15

nitrogen: added to food by fermentation, 168; in animals vs. plants, 8–10; in carnivore diet, 18; in panda's bamboo diet, 218n16; phosphorus found with, 12; umami taste and, 10–11

non-ruminants: extinction of, 230n23; likely preferred by Clovis people, 105–6, 230n23. *See also* megafaunal extinctions; proboscideans

Norbrook, David, 215n3

noscapine, 16

nose: of dog, 64–65; of human, 52, 65–67; of pig, 55, 57. *See also* olfactory receptors; orthonasal aromas; retronasal aromas

nucleotides, 9, 11

odors. *See* aromas

okapi meat, 243n6

oleogustus, 217n12

olfactory bulb, 57, 64, 72–73, 226n11

olfactory receptor codes, 73, 74, 226n12

olfactory receptors, 57, 72–73; of dog, 65; evolution of, 63–64, 66; of humans, 66, 225n10. *See also* nose

omnivores: balance of chemical elements in, 5, 9–10; flavor of meat of, 99–103, 229n17; phosphorus and, 12

On Food and Cooking (McGee), 46, 70

onions, 141–42

On the Nature of Things. See Lucretius

optimal foraging, 91, 93, 94, 95; by predatory mammals, 96

opuntia cactus fruit, 232n6

orthonasal aromas: decreased human sense of, 66; defined, 62–63; dog's nose specialized for, 65; of fermented meats and fish, 179

oxytocin, 206–7, 243n5

pacas, *94*, 95, 102, 103, 228n11

pandas, giant, 19–20, *20*, 218n16

Parmigiano-Reggiano (parmesan), 192, *198*

pastoralists, planning to find a preferred flavor, 101

Patagonia, Arizona, 80–81, 91–92, 95, 110

Patisaul, Heather, 206

Patterson, Penny, 49

pawpaw fruit, 232n7

peccaries: flavorful to hunters, 95, 96, 101, 103; giant, 88, 122; as omnivores, 99

Penicillium fungi: on bloomy-rind cheeses, 194, *198*; of Cabrales cheese, 195–96; metabolizing lactic acid, 241n9; *P. camemberti*, 194; *P. roqueforti*, 195

Pentadiplandra brazzeana, 41–42, 115–16

pepper. *See* black pepper; chili peppers; sichuan peppers

phenylthiocarbamide, 224n27

pheromones, 58–59, 225n3, 229n21

phosphorus, 12–13, 18

Physiologie du goût. See Brillat-Savarin, Jean Anthelme

pig knuckles, Cantonese black vinegar, 109

pigs: human experience of flavor compared to, 69; with meat having aroma from male pheromone, 229n21; noses of, 55, 57; sour tastes and, 163–64; truffles and, 54, 57, 59, 62, 69; wild, flavor of meat from, 100, 101

pigtail monkeys, 164

pine grosbeak, 229n18

piperine, 148, 235n13

plants: defensive chemicals of, 102–3, 106, 131–34, 153; eaten by chimpanzees, 39–40, 220nn13–14, 221n15; as medicine, 139, 234n7; with seeds storing energy in fat, 228n13; sometimes giving unpleasant flavor to meat, 102–3; spheres of chemicals on leaves of, 130, *132*, 233n2; storing energy in carbohydrates, 98

pleasure: alcohol from fermentation and, 167, 169, 180; ancient human questions about, 1–3; aromas hardwired for, 57, 58–59; brain chemicals and, 3, 6; Brillat-Savarin and, xii; central to human evolution, 24; of companionship when dining, 210; divided perspectives on, 3; food sharing and, 206–7, 208; memory of an aroma and, 75; pursued by ancient hunter-gatherers, 110; of spices, 153. *See also* displeasure

poisons. *See* toxic chemicals

predators: biological stoichiometry and, 4, 5; hunted but not always eaten, 95; musk glands of, 229n16; simple guts of, 229n17; tasting like low-fat beef, 99

primates: differences in taste receptors, 22–23; haplorhine, evolutionary changes in, 65–69. *See also* chimpanzees; gorillas; monkey meat; monkeys and sour tastes

proboscideans, 108, 110, 121–22. *See also* elephants; gomphotheres; mammoths; mastodons

processing of foods: by ancestors of humans and chimpanzees, 43–44, 46–47; by ancient humans, 32, 34–37; complex aromas and, 70–72; freeing time and energy for other pleasures, 47

pronghorn meat, 105

proteins: aromas from sulfur compounds in, 97; in muscle of cooked meat, 97; nitrogen in, 9, 10, 11; in panda's bamboo diet, 218n16

The Psychology of Flavour (Stevenson), xii–xiii

ptarmigan, 101, 102

puhadi, 142–43

putrescine, 57–58, 225n2

rancid butter aroma, of fermented herring, 179

raw, unprocessed foods, 34, 46, 48, 220n10

recent humans: Cro-Magnon, 111; culinary traditions and cuisines of, 51; defined, 32; lineages of, 51; similar taste receptors in lineages of, 51. *See also* Neanderthals

Reed, Danielle, 15

Reshef, Hager, 108–9, 110

retronasal aromas: of chemicals in fat, 99; in chimpanzees and gorillas, 67; as component of flavor, 68; of fermented meats and fish, 179, 180; increased human sense of, 66, 68, 69; loss of transverse lamina and, 66; path of exhaled breath and, 65; of truffles, 63

ribonucleotides, 11

roots: benefits from cooking of, 46, 48; defined, 219n4; fermentation of, 35, 47, 167–68, 180; giving flavor to meat, 100, 101, 102, 106, 229n17; processed to facilitate chewing, 46–47, 48; processed to release nutrients, 34

Roquefort, 195

rotten egg aroma, of fermented herring, 179

Rozin, Paul, 150–52

ruminants: flavor of meat from, 100, 102, 105; megafaunal extinction and, 230n23

Sabater Pi, Jordi, 165, 221n16

Saint Benedict, 188–89, 191, 197. *See also* Benedictine monks

salicin, 16

salt: animals' needs for, 7–8; ashes of plants used as, 130; chimpanzees' attraction to, 38; dolphins' inability to taste, 19; favoring some bacteria over others, 202; in Roman fish sauce, 239n13; washed-rind cheeses and, 196, 197, 241n10

Salt (Kurlansky), 177

salt curing of meat, 176–77

salt taste receptors, 7–8; desert mammals and, 22

Samuni, Liran, 205, 206, 243nn4–5

Sauer, Jonathan, 157

sauerkraut, 156, 164

Saul, Hayley, 137–39

savory taste. *See* umami taste

scale insects, 10

Schaal, Benoist, 135

sex, aromas hardwired for, 58–59, 225n3, 229n21

sharing food: by chimpanzees, 205–8, 209, 242n4; with human conversation, 207, 209–11

shark meat, fermented, 239n12

Sheldrake, Merlin, 56, 168

shellfish eating, 33–34, 47–48, 51

Shepherd, Gordon, xiii, 64, 68, 72, 75, 199–200, 203

Sherman, Paul, 139–41, 143

sichuan peppers, 148–49

sickness: aromas associated with, 140. *See also* food-borne illness

silica, in grassland plants, 131

silphium, 227n6

Silvertown, Jonathan, 11

sloths: holding out on islands, 89; terrible flavors of, 230n24. *See also* giant sloths

Smalley, John, 158

smeared-rind cheeses, 196

smell: differences among species, 54; evolution of, 63–64, 66–67. *See also* aromas; nose; olfactory bulb; olfactory receptors

smoking meat, 176

sniff: bipedalism and, 67; of dog, 64–65, 67

sodium. *See* salt

soft cheeses, aged, 190, 192–99, 241n8; bloomy-rind, 194, *198*; blue, 195–96 (*see also* Cabrales cheese); washed-rind, 74, 196–99, *198*, 241n8

sour beers, 156, 157, 162

sourdough bread, 157, 162

sour taste: acidity detected by, 15; animals that enjoy, 163–65; aversion of many animals for, 163; chimpanzees' attraction to, 38, 164–65; of fermented fruits, 160, 167, 169, 180, 237n6; fermented meat or fish and, 169, 172, 173, 175, 179, 180, 181, 238n9; human enjoyment of, 164; as a mystery, 15, 154–56; newborn baby's response to, 155, 234n5; safe fermentation indicated by, 162, 181; of very short fatty acids, 217n12

sour taste receptors, 155, 163, 236n1

Spence, Charles, xiii

Speth, John, 174

sphagnum bogs, 171, 238n9

spices: almendro seed, 160; ancient Roman meat recipe with, 144–45; antimicrobial activity of, 140–41, *141*, 142, 143, 145, 146, 147, 152–53; archaeological record and, 137–39; black pepper, 131, 147–48, 235n13; chemesthesis and, 147–52; culinary danger added by, 149–51; defined, 130; fetal experience of, 134–37, 233n4; in 4000-year-old proto-

curry, 235n11; hunter-gatherers'
use of garlic mustard, 138–39;
intentionally using toxic plants,
134; as medicines, 139, 234n7; not
universally used by humans, 130;
parts of plants used as, 130–31;
pleasure of food and, 147, 153; used
more in hotter regions, 143–44, *144*
starches, 14
steak tartare, 223n23
Steiner, Jacob, 233n5
stems: fermentation of, 158, 168; tubers
as a form of, 219n4
Stevenson, Richard, xii–xiii
stews: Babylonian puhadi recipe,
142–43; of hunter-gatherers about
4600 BCE, 138–39; meat of older
animals in, 104–5; Nigerian egusi
using bitter leaf, 139
Stilton, 195
stinkfish, 179, 240n14
stinking toe tree, 118–20, *120*, 128
stoichiometry, biological, 3–5; salt
and, 7
stone tools: of ancient humans, 34–35;
Clovis points, 82, 84, *85*, 90; of first
people in the Americas, 82–83,
84; used to butcher mammals,
82, 109
storing food: as aged, hard cheeses,
186, 191; by fermenting fruits and
vegetables, 168; by making cheese
from milk, 186; methods used for
meat, 169–77, 238n9
Stringer, Chris, 32
strychnine, 16, 133
sturgeon, rare, 89
Styrbæk, Klavs, xiii
sucrose, 14
sugarcane, eaten by chimpanzees, 43

sugars, 14
sulfurous cheeses, eaten during
pregnancy, 135–36
surface ripened cheeses, 194, *198*
Surovell, Todd, 229n17
sweetness: chimpanzees' choice of
fruits and, 40; energy to fuel a
larger brain and, 32–33; newborn
baby's response to, 234n5
sweet taste receptors, 14; body size of
animal and, 217n10; fruit protein
that short-circuits, 41–42; gorilla
mutation in, 41–42, 221n17; lost in
birds, 21; lost in carnivores, 18–19;
of marmosets, 22; similar between
humans and chimpanzees, 38;
similar between recent human
lineages, 51
sweet-umami receptors, in hummingbirds, 21
swifts, evolving sweet-umami
receptors, 21

Takahata, Yukio, 76
tannins, 106
tapir meat, 95
taste: as component of flavor, 6, 68;
Latin word origin, 6; preferences
of newborn babies, 233n5. *See also*
Brillat-Savarin, Jean Anthelme
taste buds, 6, 216n8
taste receptors: calcium and, 13;
evolutionary changes in, 17–23;
human efforts dulling natural
selection on, 23–24; locations of, 6,
216nn7–8; losses of, 17; phosphorus
and, 12–13; pointing animals away
from danger, 15–17; pointing
animals to needed foods, 15, 41;
signals sent to brain from, 6–7;

taste receptors (*continued*)
similar between humans and chimpanzees, 23, 37–38; similar between recent human lineages, 51; thresholds of detection, *18, 22. See also* bitter taste receptors; salt taste receptors; sour taste receptors; sweet taste receptors; umami taste receptors

"tastes like chicken," 97

teeth: DNA from fossils of, 51, 223n26; smaller in humans than chimpanzees, 32, 33, 49–50; stones serving in place of, 35

temperature receptors, 148, 235n13

termites, eaten by our ancestors, 43

terroir of meat, 101

thyme, 132–33, 134, 233n2

thyme basil, 133, 233n3

tofuyou, 201–2

tools: of ancient humans, 34–35, 77–78; of chimpanzees, xiv, 26–29, 29, 30, 32, 35, 37, 41; of common ancestor of humans and chimpanzees, 43–44, 46–47; hunting and, xiv–xv; used by our ancestors to seek flavor, 23–24. *See also* stone tools

Tordoff, Michael, 12–13, 215n1

toxic chemicals: of bacteria in decaying meat, 171; bitter, 16–17, 133–34, 217nn13–14; in plants, 100, 131, 133–34

transverse lamina, 66–67

tree leaves, and flavor of meat, 102–3, 106

tree shrews, broken TRPV1 gene of, 235n13

trimethylamine, 136

TRPA1 receptor, causing tingling, 148–49

TRPM8 cold receptor, 148

TRPV1 heat receptor, 148, 235n13

truffles, 52, 54–57, 60–62, 70

Tuber, 55, 224n1

tubers, 46, 219n4, 223n22

Tyone, Mary, 179

umami taste, 10–12; of aged soft cheeses, 193; of avocados, 96; of fermented foods, 168, 179, 180; newborn baby's response to, 234n5; of salted meat, 177; of shellfish, 47

umami taste receptors, 12; broken in giant pandas, 20; broken in sea lions and dolphins, 19; evolved to detect sugars, 21; similar between humans and chimpanzees, 38; similar between recent human lineages, 51

Ungar, Peter, 50

Vaillant, François Le, 109

van Zonneveld, Maarten, 126, 127–28

vinegar: aroma of fermented herring and, 179; sour taste of, 164

vitamin B$_{12}$, added by fermentation, 35, 168

vitamin C: not synthesized by humans, 216n5; sour taste of, 155

vomiting: aroma associated with, 140; triggered by bitter compound, 17

Waorani people of Ecuador, 103, *104*

Warren, Robert, 126–27

warthog, delicious, 101, 229n19

wasabi, 148

washed-rind cheeses, 74, 196–99, *198*, 241n8

Wejendorp, Kim, 104

wet fermentation of meat and fish, 177–79

wild boars, meat of, 101, 229n21
Williams, William Carlos, 77
wine: fermentation of grapes and,
237n6; learning to discriminate,
200
wine experts, 74, 75, 226n13
Wittig, Roman, 205–6, 242nn3–4
woolly mammoths: causes of extinction,
228n8; in Europe, 89; meat of, 108.
See also mammoths

Wrangham, Richard, 36–37, 49, 220n11,
221n15

yeasts: alcohol produced by, 156, 166,
167; bacteria competing with, 166;
carried to sugar sources by insects,
237n4; fermentation by, 156–57, 160,
167, 237n6; metabolism of, 166

Zimmerman, Andrew, 118